MODEL BUILDING AND SUPER DETAILING

MODEL BUILDING AND SUPER DETAILING

Detailing Techniques Including 3D Printing

David Ashwood

&

The Market Deeping Model Railway Club, CIO

PEN & SWORD TRANSPORT

AN IMPRINT OF PEN & SWORD BOOKS LTD.
YORKSHIRE – PHILADELPHIA

First published in Great Britain in 2024 by
Pen and Sword Transport
An imprint of
Pen & Sword Books Ltd.
Yorkshire - Philadelphia

Copyright © David Ashwood, 2024

ISBN 978 1 39909 488 7

The right of David Ashwood to be identified as author of this work has been asserted by him in accordance with the Copyright, Designs and Patents Act 1988.

A CIP catalogue record for this book is available from the British Library.

All rights reserved. No part of this book may be reproduced or transmitted in any form or by any means, electronic or mechanical including photocopying, recording or by any information storage and retrieval system, without permission from the Publisher in writing.

Typeset in 11.5/14 Palatino
by SJmagic DESIGN SERVICES, India.

Printed and bound in China by 1010 Printing International Limited

Pen & Sword Books Ltd incorporates the imprints of Pen & Sword Books Archaeology, Atlas, Aviation, Battleground, Discovery, Family History, History, Maritime, Military, Naval, Politics, Railways, Select, Transport, True Crime, Fiction, Frontline Books, Leo Cooper, Praetorian Press, Seaforth Publishing, Wharncliffe and White Owl.

For a complete list of Pen & Sword titles please contact

PEN & SWORD BOOKS LIMITED
George House, Units 12 & 13, Beevor Street, Off Pontefract Road,
Barnsley, South Yorkshire, S71 1HN, England
E-mail: enquiries@pen-and-sword.co.uk
Website: www.pen-and-sword.co.uk

or

PEN AND SWORD BOOKS
1950 Lawrence Rd, Havertown, PA 19083, USA
E-mail: uspen-and-sword@casematepublishers.com
Website: www.penandswordbooks.com

Contents

1 Introduction 7

2 The Evolution of Detail 11

3 Dressing the Stage 16

4 Military Matters 27

5 Fore- and Background... 31

6 Water Features: Seas, Rivers and Puddles 35

7 Vignettes.. 42

8 Railways Within a Scene... 55

9 Detailing Rolling Stock 60

10 Building Your Own Rolling Stock 67

11 Entry-level 3D Printing 79

12 3D Resin Printing... 89

13 Computer-aided Design 97

Why we 'Do' Model Railways 106

1
Introduction

New for old, but almost identical in design and technology. In modelling, taking detailing decisions can result in the learning of a new up-to-date skill, but at times the old way can still be the best way. This is BR (SR) at Winchfield Hants on 12 September 1956. The S&T crew have the new gantry ready for use with just the signal arms to mount, creating a temporarily cluttered driver's-eye view. Beyond this point the M3 motorway now sweeps through the woodland and over the railway. The SR utility van DS10 that can be glimpsed is a shorter two-axle version of the bogie van used in the Tri-ang detailing project later in this book. (*Online Picture Archive AND-M326-1*)

The common man is not concerned about the passage of time, the man of talent is driven by it.

Arthur Schopenhauer

The Market Deeping Model Railway Club (MDMRC) was formed in 1976, the year Steve Wozniak designed and released the Apple-1 computer. This club, in common with all of its ilk, is a thriving social ecosystem of like-minded railway modellers, with a desire to learn, share, specialise and display the end results to the public at exhibitions. The pictures in this book are taken from a combination of Club, exhibition and home layouts to represent the broad spectrum of this hobby, from the smallest static displays to larger longer-term builds.

Like so many that exist in the United Kingdom and elsewhere, this club has an interesting mix of contributing talents, a can-do attitude and a structure of 'officers' that can cope with planning and budgeting through to handling the most unexpected of circumstances. Then disaster struck on the night of 17 May 2019.

The Annual Model Railway Show was set up at a school in the old limestone town of Stamford, Lincolnshire, on that Friday evening. Layouts were set up and tested, 'position one' rolling stock put in place. Traders arrived and their wares were set out ready to sell on their displays.

During the night the school was broken into, and an extensive spree of vandalism occurred.

At 6:30 am the school caretaker ,accompanied by author and his wife, opened up to discover most of the contents had been comprehensively reduced to matchwood. Our own layouts, those of fellow clubs and the traders' stalls were all demolished.

It was the last thing you would expect to see. In some cases, twenty-five years of work were gone; kindly men over seventy were in tears. The Club and all others present experienced shock, anger and disbelief. That Saturday afternoon was spent with brooms and dustbins, and we wondered just how we would cope with the damage and loss of Club earnings. We decided to set up a £500 'Just Giving' request online to try offset losses.

During the following week, news of the vandalism was flashed around the world, seizing the common imagination. Club members appeared on television and radio. Our one operable locomotive, taken from a raffle prize, was set up on our Club's test track to show some background movement.

Within a day, members of the railway modelling fraternity, wargamers, the general public, people with fond memories of their grandfathers' past, provided kind words,

Looking like a post riot scene rather than an exhibition. This is what greeted the Club in the main hall. The result of an episode of comprehensive demolition. Always make sure your own layout and stock is insured adequately, either with house insurance or as a separate specialist policy.

and donations of all kinds poured in. From Miniatur Wonderland of Hamburg and Sir Rod Stewart through to children's pocket money and a lady from Japan apologising for her English. It should be realised that there was not only the financial investment, but the irreplaceable time and devotion of past and older members, ideas, aspirations, discovery – all had been lost.

The Market Deeping Club formed a charity to process those donations appropriately, curate a historical collection of assets representing the evolution of the hobby in Britain and elsewhere, promote model clubs for local children and assist other local good causes. Good can come from bad, eventually. Our share of profit from this book goes directly back into the charity fund.

This publication is aimed at sharing Club members' experience and skills involved in constructing and improving models, which is an integral part of most display layouts. There are followers, collectors, specialists and generalists. Above all else, the modeller is someone who can project their inner self into a tangible object.

A part of the literal rebuilding process within the Club has seen a return to the roots of the hobby. How does one repair a broken object, or change it, or take the brave decision to replace it? This could be a fence, a platform, rolling stock or a whole baseboard. We hate to see loss of the investment of years, indeed any waste. Some layouts were written off, their assets stripped and placed into scrap boxes for eventual reuse.

What has emerged from this smelting furnace is the willingness to adopt new techniques to supplement the old, such as 3D printing to layer detail over existing models, laser-cut components to add to a scene. Newer materials and techniques such as static grass, resin water and oven-baked mouldings have replaced dyed tea leaves, paint and varnish or whittled balsa wood in many, but not all, areas.

This book goes into deeper detail than the previous two. We are now into the realms of finessing models and have tried to give examples of many different ways a modeller can progress.

Books such as this cannot exist in pure isolation, they serve as a launch pad for greater things: read periodical magazines, use the library service, search the internet, visit model shows, ask questions and see what can be done. Model shop owners and exhibitors get lonely, they love to talk. Above all, enjoy yourself in discovering the pleasure of a perfect little world where the trains will always run on time.

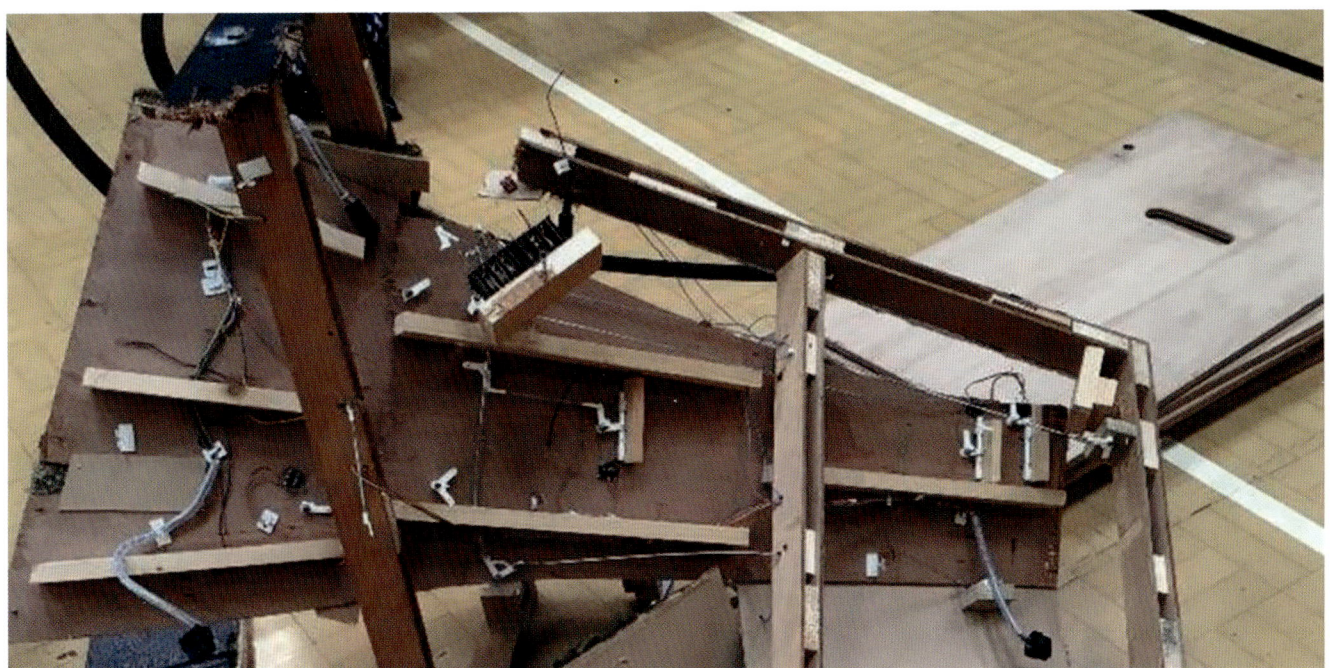

Fragments of dreams. A dislocated baseboard, broken scenery, servos, wiring. Much had to be written off completely, some could be salvaged. Normally an established club would wear out a layout over long service at exhibitions. The replacement cycle is normally planned and budgeted for. Suddenly our Club had gaps to fill – from bad could come good.

Out thanks go to Pen & Sword Books for proposing this third book of the series, with the aim of presenting another facet of the modelling hobby beyond the military and technical. This extensive coverage is sure to tempt anyone who has looked twice longingly at a museum exhibit or a model kit.

Thank you to Alan Hancock and Peter Davies for proofreading, Brian Norris for his skills with 3D printing and members of the Market Deeping Club for their kind assistance and willingness to share and to the layout exhibitors and operators that kindly allowed the author to take so many detailed photographs.

Club layout Canons Cross (OO) with film director Pip Swallow planning camera shots. Viewed from a 3D printed windmill, through rescued and refurbished model houses along Watling Street with the railway travelling through the tunnel below. To the left is a 3D printed model village and martial arts dojo by degree students from the University of Hertfordshire. Four metres beyond the extensively modified Metcalfe church and streetlamps is the terminus of the station. This project represented the Covid lockdown rebuilding of the Market Deeping Club's hopes and aspirations. (*Author*)

2
The Evolution of Detail

Everything feels a little ramshackle and out of square, but it still serves its original purpose. Before the widespread availability of accessories and materials suitable for detailing there was a lot of inspired invention. Looe station terminus on 16 August 1966. Swindon cross-country Class 120 duo W51573 and W51582 from the Laira 84A allocation. These blue square coupling code units survived into the mid 1980s, often being used for parcels traffic. (*Online Transport Archive – Meredith 627-5*)

Early Days

The Model Railway Club ('The MRC') of London was established in 1910 and is the oldest such club in the world. It is the author's recollection of the shows at London's Westminster Hall in the 1970s that has led to fifty years of keen model-making. More recently, seeing the MRC 2mm finescale Copenhagen Fields model at the Alexandra Palace triggered dormant ambition. Memories of the sprawling King's Cross approaches of that model are what prompted

Above: Post Second World War scratch building by an ex POW from 'Klim' milk tins. These O gauge models are on Hornby and Märklin clockwork chassis and were recently donated to the Club. They represent early artisan attempts to break away from the normal offerings by the toy companies. We will show some more up-to-date scratch building later in this book. (*Author*)

Left: The pre Second World War Hornby tinplate was ideal in the eyes of the youngster with imagination. It was a real railway under personal control. (*Author*)

Right: To find a midpoint between the kit modeller and the ready-to-run, Tri-ang in the 1960s produced CKD ('Complete Knocked Down') build-your-own models. A cynic would say this was a way of increasing sales while reducing factory production costs. Perhaps, but it also met a need for that endorphin rush feeling of success and ownership that comes when you apply yourself to making a model. (*Author*)

Below: Compare the Hornby tinplate with the Metcalfe signal box kit on The Priory layout in N gauge. So much smaller yet with far more detail. (*Author*)

Above: Overall a child prior to the mid 1960s had to use more imagination over substance where model railways were concerned. Here is a 1950's child's eye view of die-cast and tinplate Hornby Dublo from the author's collection. (*Author*)

Two- and three-rail representations of the Hornby Dublo Travelling Post Office set. Competition with the injection mouldings from Tri-ang led to a somewhat clunky transition to plastics by Hornby, converting existing models at times rather than replacing them, as they were wrong-footed in the marketplace. (*Author*)

In some ways a cruder operation (a mechanical- rather than solenoid-driven mail catcher) but associated with better infrastructure. The Tri-ang models made inroads into the Hornby market. Such competition drove diversity, and over time both innovation and quality. (*Author*)

the author take on Euston Station in the past few years.

Industrial engineering models have been around for a long time, smaller versions of the prototype made by talented individuals or apprentices. It is the mass-produced model that brought the toy train into the hands of children in the twentieth century. Such companies as Basset Lowke, Lionel, Hornby and Märklin evoke mental images of brightly coloured tinplate. These companies brought standards of gauge and scale into the marketplace which meant that the aspiring modeller could concentrate on buildings and scenery instead of having to be an expert at making track and rolling stock from scratch.

While it is tempting to see these early years as ones of crude representations of the prototype, there were groups and individuals who sought to bring more realism into the hobby. Woodworking, engineering and artistry all

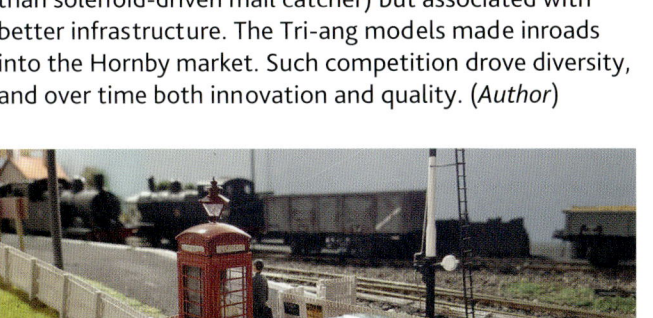

Above left: To complement the real-life image of the chapter header, compare a modern level of detail with the Butterwick station access road in O gauge. (*Author*)

Above right: Or with that of the station throat of the Bantry layout in OO from Sleaford MRC. (*Author*)

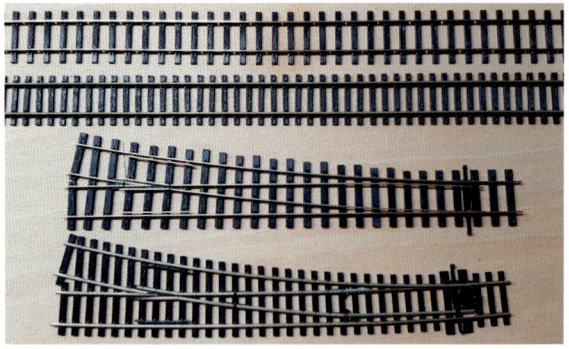 

Above left: Occasionally mass-produced items catch up with the hand-crafted. When demand is seen to exist, and the market will support a premium price, you get a jump that pleases the modeller. For example, Peco have produced code 75 (finescale) flat-bottomed rail for some time, but recently introduced UK sleeper spacing with bullhead rail. It is only when you see them side by side that you realise how visually pleasing this is. Turnouts with nine fewer sleepers look correct in UK older prototype models. The downside is that coarser-flanged stock will not work on these rails, and merging rail types can be expensive. (*Author*)

Above right: Another innovation is the Unifrog turnout. You can power the 'V' or leave it isolated as you wish. A boon for DCC (digital command control) modellers, as well as those who like to 'juice their frogs' with a micro switched power supply and those following traditional wiring concepts. (*Author*)

Many preserved railways have made good use of second-hand bullhead rail, but in many places on the main line network it still hangs on today. Here seen at Sandy station on the East Coast Main Line is a siding laid with wooden sleepers, iron track chairs and clips replacing what would once have been oak blocks. (*Author*)

came to the fore as third rail gave way to stud contacts and 'coarse scale' became 'finescale'.

New constructional materials and adhesives evolved, popular kits from wood, card and plastic began to fill the market and the model offerings became smaller in size and no longer just within the grasp of the richer members of society.

In the 1950s and 60s it seemed most schools and churches had their own modelling clubs for the youngsters (when they were not trainspotting), in Saturday morning cinema club or Sunday school. Hornby Dublo and Tri-ang battled it out until Binns Road was eventually subsumed by Lines Brothers.

A GWR (Great Western Railway) bridge rail from the scrap heap at Southall shed was sliced for GWRPG (Great Western Railway Preservation Group) members. These formed the core of Brunel's broad gauge track formation spiked direct to longitudinal sleepers. They are often reused as fence posts and sign supports. We find that scrap Hornby Dublo track works very well for such purposes. (*Author*)

Bullhead rail length reused as a parking guide in Peterborough station car park. If you have scrap or substandard track, make good use of it in your extra details. Sleepers stack up, offcuts become trackside rusty scrap, track fence posts become resistant to accidental knocks. (*Author*)

Papier mâché, dyed sawdust, tea leaves, balsa and wool formed the basis for scenic extravaganzas alongside town and station models.

Today the absolute scale, the true colour, the feeling of reality is all owed to these inadvertent pioneers and the techniques they piloted. In parallel they also developed a modelling market into which companies keenly infilled.

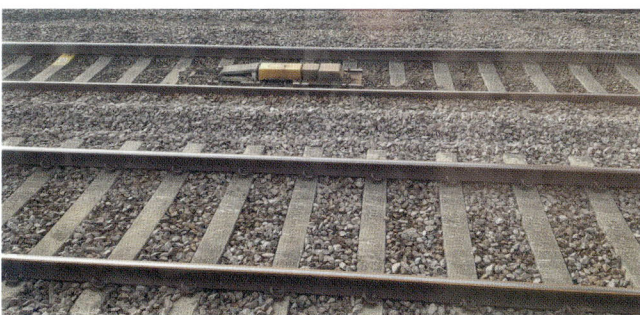

The main running lines heading to King's Cross have Dowmac concrete sleepers in closer spacing and flat-bottomed rail retained by the twisted Pandarol clips. Note the AWS (Automatic Warning System) ramp in the background. (*Author*)

Top, above and below: A model railway is whatever you want to make of it. Challenges can be diversified and it can vary personally from basic to complex as desired. Club member John Harrison has fun with a layout where all scenery is made by recycling domestic waste. Card, bottles, lids, all find a place. At the same time he curates Unserstadt, originally made by Brian and Ann Silbey in N gauge, which is used for exhibition purposes. (*Author*)

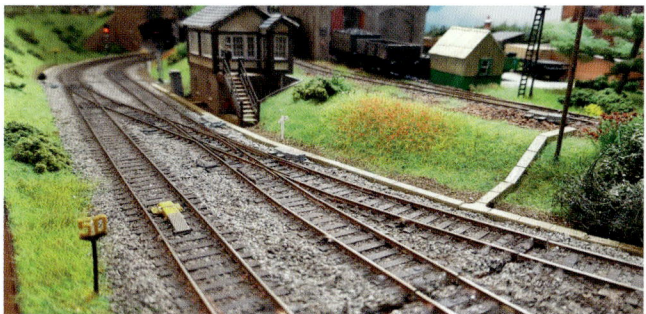

Similar track detail seen on the OO Witham layout of member Martin Reynolds. These little flourishes add to the complexity and believability of a model. (*Author*)

Queen greets *Duchess*. Leftmost is R3855 46211 *Queen Maud*, an elegant, finely balanced locomotive from Hornby. This had just been repaired for a friend after the box was dropped and the motor to final drive universal joint unslotted itself. Compare to the venerable Hornby Dublo three-rail die-cast EDL12 46232 *Duchess of Montrose* – sparky, noisy, smells of geraniums. Proven by the author to be nearly indestructible in childhood play, I recall this one falling off the table a few times without damage. The quality of detail has improved hugely, but naturally at the expense of robustness. (*Author*)

3
Dressing the Stage

Class 122 'Bubble Car' W550003 single carriage unit sits at Stratford-upon-Avon on 30 May 1975. The station still has tended flower beds with whitewashed stones surrounding. The brown and cream enamel signage of the previous era is looking rather careworn. With judicious set dressing your stage can represent quite a spread of years against which you can change the period of rolling stock portrayed. After all, who would expect rail blue and the previous era to look quite so good? Built in January 1958 this GRC&W (Gloucester Railway Carriage & Wagon Company) DMU is currently preserved and running on the Gloucestershire and Warwickshire Railway. (*D Mulquin/GWRPG*)

All your little world is a stage. Great fun can be had by tailoring kits and using printed/embossed sheet materials to make your own accurate representations for location and time period portrayal. At all times, if you think theatrically, you will impart the overall feeling of the years passing. In your own 'stage dressing' give thought to the following.

- **Research**. Take time to look at old images of the location and period. It is amazing how many aspects of life get forgotten, even over a period of ten years, let alone looking back through generations.
- **Consistency**. Stick to your guns, resist mission creep and keep the original requirements in mind. Even if you rather

liked that Shelvoke dustcart, it wasn't around in 1923.

- **Colouration**. Colour complexity increased over time as better pigment mixing and surface finishes became available. For example, the LNWR (London and North Western Railway) stuck to very basic colours on stations, requiring a set ratio of pigment to white base paint. Long lasting but rather dull. Compare to the livery on modern carriage sets and regional stations, let alone the colours available to the householder. Try to stick to the colours of your period, it will look right.
- **Extras**. Your people, their jobs and their style of dress reflect your hard work on infrastructure. Similarly, using accurate road transport types of the period, from horse-drawn through internal combustion engine and now electric will pitch the mindset into the period portrayed.
- **Minutiae**. Little things count – they'll trigger an empathetic response in the viewer of your model. That is why time is spent on manhole covers and meter boxes. Think of shop window dressing and advertising posters to set your time period.

Our 1901 scene is set during the construction process, one shop as yet unoccupied.

- Estate roads are rolled clay or sand surfaced with sprayed tar. Not very resistant but would work with the lighter weights of the horse-drawn traffic. Higher traffic roads would often have granite cobbles or tar wood blocks as the wear surface. Here painted with burnt umber acrylic pulled through beige with a broad brush.
- Consistent foul water drainage was a Victorian innovation (the Romans can be ignored for our purposes). Cast-iron gratings

1901 townscape: a bright new future. Many towns and cities saw unprecedented growth during the Victorian era as the rural population moved to the urban centres aided by the railways. Plots of land were swallowed piecemeal, larger estates and houses demolished to make way for regimented rows of terraces and semi-detached villas. The local vernacular was replaced by a standardised catalogue building style. Houses of this age reflected aspirations as each was designed for a single family, often to rent or lease (owner-occupation was at this time something of a minority accomplishment).

18 • MODEL BUILDING AND SUPER DETAILING

1940 townscape: wartime duress. Storm clouds gather over Europe, our houses and shops are now thirty-nine years old and beginning to show some diversity. Trees and shrubs are larger, some technology has reached the scene. Overall things are still recognisable from the Victorian portrayal as the pace of change has been moderate.

and manhole covers were often made at a local foundry and many regional variations exist. Here we used cutout examples printed from screen captures and personal photographs.
- Gardens are freshly planted, everything looks small and neat.
- The same is true of paving, newly laid and tidy. Images printed from free-to-use textures online.
- Our café establishes a food use type of occupancy with the local authority, which would need council permission to alter.
- Posters, signage and paintwork are all within simple variations of available pigments of the turn of the century. In the absence of a colourful enamel advertising sign, it was likely to be a playbill or locally printed information.
- Windows normally had thick net curtains for privacy and heavy curtains to blackout and stop draughts. We have used kitchen towel for this netting texture.
- Whole areas would look homogeneous in colour: design, fencing, walling and gates. Our Metcalfe kits here are very much out of the box.
- Most noticeable in our 1940 scene is perhaps the anti-blast protection on the windows. Such tape reduced the inward shattering of glass in the event of a bomb explosion.
- Gardens are wilder than they would have been previously. The 'dig for victory' campaign meant that they were often dug up for food production. Members of the family would have been absent so less time and energy could be lavished.
- Paving began to crack up – just the simple use of an HB pencil here adds to the scene.
- The railings and garden gates have been removed for salvage. Pre-war churchyards, walls and gates would have been full of wrought ironwork and castings. During and after the war, just the stubs would remain. On the O-gauge Butterwick layout,

DRESSING THE STAGE • 19

1969 townscape: down at heel. Victorian housing became effectively life expired without refurbishment by the 1970s. Lead and ironwork sprang leaks, there was a lack of maintenance since the war as landlords struggled to cover the requirements to modify the explosive Ascot water heater, change cookers for North Sea gas as well as repairing rotting window casements. Wartime damage was often covered up and conveniently forgotten about. The backstory here is that our example lost part of its garden wall, fancy window glass and a shop front to blast damage.

walls have been 3D printed with rusted iron stubs where railings would have been.
- A new manhole for a water hydrant has appeared. The garden wall painted cross with EWS (Emergency Water Supply) marks where this is placed.
- A telephone box has been located on the vacant lot. The GPO (General Post Office) had to negotiate for street positioning, often taking the space that would have been a front garden plot but is now for a shop. Here the concrete apron of the shop would be private property, hence it has not been paved.
- Although many of the photographs before 1950 would have been black and white, some research (and a long memory) helps to set the scene. In the 1930s through 1950s many urban district councils took advantage of a cheaper supply of red granite chippings laid on top of a tar layer. The roads weathered to a gentle pink as a result. Three colours of acrylic – raw umber, burnt sienna and buff titanium – have been blobbed then lightly brushed through each other in a linear fashion giving the gentle gradation that would have existed.

By 1969

- An extra road manhole has appeared as foul water and rainwater have been separated.
- Our once little tree is now large and inappropriate for the setting. Many gardens suffered going from beyond maturity to outsize and clutter by this time period.
- The nearer house has covered up cracking from bomb damage with plaster and at this late 60s juncture has a lovely application

20 • MODEL BUILDING AND SUPER DETAILING

1999 Townscape: resurgam! Anything Victorian is becoming sought after and trendy once again. Those houses with a decent build quality and that have survived redevelopment are now being suitably refurbished. Doors and bannisters are being unveiled from under hardboard internally (reversing the DIY aspiration of the BBC's Barry Bucknell in the 1960s), guttering and roofing are being replaced externally. These kits have been extensively embellished and the overall feel is a movement from Victorian simplicity to the complexity of modern life.

of garish paint to supplement mould and rising damp. It has also been subdivided by the landlord and a dormer squeezed into the loft space.
- Bins out front, not taken to the rear any more. Also note the start of on-garden parking by the moped.
- External TV aerials are appearing. Twisted multicore wire for OO and soldered staples to wire for O.
- The paving has seen some renovation. Cars often would park with wheels on the pavement, so urban councils countered the break-up of paving by using asphalt near the road.
- The GPO has installed manhole covers (iron surrounds and concrete middles) for the telephone service. This area is better than many – no poles and overhead wires in place.
- Hydrant signage with the black H on a yellow background and a yellow painted cover has appeared.
- The road is now tarmac blacktopped and beginning to see patch repairs where utility companies have been digging. Payne's grey acrylic was used here – roads are often portrayed as jet black, but in reality they weather quickly to greys.
- Advertising is more professional and prominently sited with national companies making neater displays.
- The Golden Egg and Mac Markets were chain innovators of their time, occupying many streets and attracting a clientele away from the greasy spoon and corner shop. Vesta noodles and curry anyone?

In 1999

- The urban front gardens have lost almost all their greenery, being bricked over as hard standings for parking and rubbish.
- Say goodbye to the old tin dustbin, it's black bag pickup day today. Side access of houses is now blocked by doors and gates to prevent trespass.
- The flats are official and now have a purpose-built dual entrance. There is a dropped curb for the house with the posh internal shutters

and modern roof extension. Chimney pots have either been removed or have become small vents for gas installations.
- The telephone box is no more. After becoming an unofficial public convenience it has now been sold off, leaving its floor mark.
- Technology has progressed. We now have a house alarm and external white utility meter boxes. The short lived BSkyB 'Squarial' can be seen on a chimney, and an analogue circular white satellite dish to pick up foreign language broadcasts is above the coffee shop. (These dishes are generally south-facing in the UK, here made from card cut out with a staple bent for the LNB collector arm.) New water stopcocks are placed in the circular holes in the nice new block paving.
- Advertising is now to the forefront with an illuminated poster setup, yet the convenience shop has no inward visibility (window from a stretched OO 3D print of a Georgian example!). Costly Coffee carries on the food retail dreams into the twenty-first century.
- Roads are more worn and patched, using tarmac texture prints; gutters have linear 'French' drains rather than the cast iron covers of old.
- A parking warden is patrolling the new yellow line restriction. Will the lay preacher get an answer from the UPVC door he so admires?

In goes the water hydrant wall sign at the end of a pair of push-to-separate tweezers. Even researching the meanings of these signs and getting it accurate is a small victory in a build project. Top number is the hydrant size in mm and the lower shows it is 2 metres away in the 1999 set. The internet search engine is a willing friend when it comes to those little extra details.

You will spend much of your time looking at your model from above, especially if your operating area is at the rear, such as on an exhibition layout. Making sure it feels right and has details that perhaps can only be seen from certain angles will allow visual reveals that are rewarding. Here is 1969 from above.

On detachment. Some of the models from this book series in a supporting role. If you build models of local scenes they can potentially be used on loan to historical societies or for lectures, as well as appearing on a layout. (*Author*)

Emergency Services

Often as you travel by train looking out of the window you witness some sort of drama unfolding, from a minor event to the major newsworthy happening. The author remembers travelling to college on the London Underground between Royal Oak and Paddington at 7:00 am on 23 November 1983 and seeing 50 041 Bulwark on its side and 606 tons of sleeper train strewn over the approaches. On return later that day the rail cranes and recovery crews were in attendance leaving an unforgettable impression on me. If you wish to find a report of a British railway incident for your prototype area you can visit https://www.railwaysarchive.co.uk/.

You can model indirect situations as well, since the railway is passing through city, town and village. Floods, fires and misdemeanours all serve to provide vignettes. Setting a realistic level of reaction is an art. Some exhibition examples witnessed appear to be a sea of flashing blue lights as the entire emergency service of a county turns up to a small car collision, thus distracting the eye from other great details of the layout.

The scenic section of Watling Street built specifically for the short film *Dream Big* contained a burning building as a part of the scripted requirement. A pump escape has unrolled its ladder as smoke drifts out of the damaged roof. A small heating smoke generator with aromatic wood fire aroma oil plus flickering LEDs create the scene. At the rear of the houses which represented the front of baseboard we had a decision to make. Unless there is major peril, local people tend to carry on with what they are doing. Rubber-necking voyeurs often seem to come from further afield. Therefore, it has been remarked that it looks normal to the viewer that Alf has not left his deckchair despite all that's going on next door.

Above and below: Images from Greeley Model Railroad Museum in Colorado showing emergency services in action, US style. Remote forest wildfires require bowsers and pumps. In towns the EMS ambulance service is often covered by the Fire Department as witnessed here in this view of a condo fire. (*Paul Ijs MDMRC*)

DRESSING THE STAGE • 23

A dwelling fully involved and with the local fire service in attendance. A two-alarm fire of the Coachella Valley MRR layout. For realism always take the working practices of the country you model into account. For the USA a 'Hook and Ladder' company would normally vent the house roof to prevent backdraft before a charged line has been deployed. Hence here plenty of dark smoke, most rooms ablaze and a hole in the roof. (*Martin Reynolds MDMRC*)

The Club layout of Euston in 1875. A pump of the then Metropolitan Fire Brigade, with chief Captain Eyre Massey Shaw on board, gallops north up Seymour Street (now Eversholt Street) past Euston Square Gardens. The impressive façade of the active Euston Square fire station dates from 1902 so is too late for this layout. Often giving a nod to something through the use of accessories works just as well as a building. Below is the less glamorous service, that of the funeral director. Heading along Euston Square towards St Pancras is a four-carriage cortège. There are many figures and vehicles available. (*Author*)

Left: The Cleveland MRC Guisborough layout has a great scene with a casualty evacuation taking place, while concerned onlookers pay attention. The ambulance has a single blue flashing light which is just as effective in this scene as the whole arrays that can be seen in more modern representations. Adding a story with the figures and their actions creates a depth to your layout scenes.

Below: This linear model by club member Barrie Church shows that the footprint of a model need not be large. Here we see a derailment rescue in a deep cutting. This is based on a real incident where an occupation bridge had been removed but someone drove through the wall onto the line. (*Author*)

Sport and Leisure

The model railway has a hypothetical population. As with their 1:1-scale counterparts people need to have some fun. Catering for their requirements can infill odd-shaped areas on a layout as well as being an interesting challenge.

While it can feel a bit of an afterthought, the same often seems true of town planners. Land that has bad building foundation, awkward shapes or a historical association can end up being used for recreation. If you do the same on a layout it works in just the same manner as reality.

You can buy backscenes containing cinema buildings, through to figures performing sporting activities off the shelf nowadays, or just build parkland. It is for you to provision the good denizens of 'Yourville' with their requirements.

Perhaps it's time to get the old Subbuteo football set out of the loft?

Right: In stark contrast to the USA, a more genteel pursuit with lawn bowls being portrayed on the Club Canons Cross layout. They seem to play their ends happily despite the frequent electric railway service in the background. A series of small board extensions included a different vignette on each one. (*Author*)

Below: A floodlit baseball diamond of Cordes Field at Coachella Valley MRC, where someone has brought their John Deere tractor to watch the game. A popcorn shack, ice-cream vendor and burger restaurant complete the ensemble. (*Martin Reynolds MDMRC*)

Above and right: Variations of the American dream – the drive-in movie theatre. With both scenarios they are showing real movies related to the railroad on mini screens. A useful retirement for old video or tablet equipment. Above is the San Diego Railroad Museum version and right that of the Coachella Valley Model Railroaders. (*Martin Reynolds MDMRC*)

Inclusivity, Ethnicity and Creed

This comes as a challenge to many modellers, especially those modelling more recent times, often due to the availability of suitable figures and buildings.

If you model a Victorian or earlier twentieth-century layout it would be peopled by a largely homogeneous population and a male/female dress code recognisable as belonging to that era. People tended to be regionalised within a national context although the railway was beginning to shake that up by giving ease of access to the masses.

As time has progressed, multiethnicity has emerged. This can be loosely described as any movement of creed, colour and religion away from a traditional core definition. Yet most model railways witnessed on display in the UK have figures set in the mid 50s or 60s (such as the Airfix/Dapol passengers and station staff). Pre-painted items often in HO scale from France, Germany and the USA have some limited diversity.

It is quite difficult to create a crowd scene – it is expensive and time consuming. If you model an inner city from the 1950's onwards you ideally should consider thinking more about the figures used and how they can be adapted. The layout and images portrayed therein would benefit in that the population would be accurate and recognisable.

Many layouts have a church, few have a chapel (Methodist, Wesleyan or Congregationalist). As you move further from the core provision by suppliers the less likely it is that the building will be portrayed. For example, a modern GWR main line model of Southall would benefit from the Sikh gurdwara sitting in the old goods yard.

Much the same applies with the physical population. For the Club Euston 1875 layout we are portraying ladies in crinolines and

Above left: The concourse at the end of the arrival platforms of Euston. The formal dress code immediately labels the period. (*Author*)

Above right: A wedding ceremony has completed, and the happy couple are standing outside the Spanish church. Very much a regional view of Southern California desert pueblo areas near San Diego. (*Martin Reynolds MDMRC*)

A wonderful representation by Tunnel Lane Models of Miranda, the star of the film *Dream Big*, in which the Club Canons Cross layout appeared. (*Author*)

gentlemen in their top hats, instantly recognisable. The Sultan of Zanzibar has not yet arrived at the station to ratify the anti-slavery treaty – it feels very much like a scene from Dickens.

At the other end of the telescope, we are making a 1990s model of the South African Railway (SAR) in the North Cape region. This was done to make good use of donated rolling stock and is a mind-broadening experience. All research is be based on web images and YouTube videos as none of us involved have visited the country (good for the carbon footprint). The national population needs to be portrayed accurately and with consideration, based on remote research. We are not trying to make a tourist brochure, just represent things on the ground in a model as accurately as we can.

So, we encourage you, dear modeller: diversify your population, think outside the traditional modelling box.

Actress Rhoda Ofori-Attah between takes on a cold set with our layout behind. (*Author*)

Above left and above right: Views of the commercial area of the Club South African HO layout. The coke delivery man is a repaint of the Merit Accessories postman and the forklift rejigs the Faller model. (*Author*)

4

Military Matters

Lost to enemy action? Actually, a mishap. This view of the Kent & East Sussex Railway is possibly on the long embankment near the Eight Bells at Salehurst. With A1X 'Terrier' number 3 Bodiam passing an unidentified derailed sister on 23 April 1949. (Thought to be 32640 on BR allocation since the Number 5 resident locomotive *Rolvenden* was long scrapped by that time.) If you have cherished locomotives or stock that will no longer run, you can still make good use of them in vignettes on your layout. (*Online Transport Archive ADav-MO27-14*)

The Military

Many modellers incorporate the military into their layouts as the context, as traffic passing through or legacy features to include. Apart from the added interest it also opens the door to a huge area of pre-built models and figures over a wide variety of scales.

Within the Market Deeping member and Club layouts we have taken good advantage of this. Euston has a colourful honour guard forming up on a platform for the Sultan of

Zanzibar. Witham runs a train of warflats loaded with contemporary tanks which entertains at exhibitions. Butterwick uses an abandoned pillbox. Andover Road is based on the Micheldever RAF reserve fuel depot.

Something not to ignore are the techniques, available scenic products and online resources of the wargaming community, whether related to a Panzer diorama of the Second World War, space marine incursion for Warhammer 40,000 or ships of the line. It is good to look at disparate web-based resources and see how things are done in images and YouTube guides or bite-size TikTok videos.

The great thing about these mini fortifications is that they can fit into nearly any terrain or feature on your layout and become a focus of interest. This concrete embrasure is in the sand dune front at Gibraltar Point near Skegness, weathering and blending in with the fullness of time. (*Author*)

There are many military artefacts in Britain that lend themselves to model railway use. The smallest defensive strongpoint is really that of the machine-gun nest or holdfast, commonly referred to as a pillbox. Here is an example of a square infantry type 26 pillbox built to defend the approaches of the North Staffordshire Railway bridge over the B5331 at Lake Rudyard near Leek. Admittedly a double brick skin would just about stop a standard bullet – for anything of larger calibre the defensive capability would be questionable. See the Defence of Britain database for numerous examples: https://archaeologydataservice.ac.uk/archives/view/dob/. (*Author*)

Abandoned and weathered by the late 1960s, the remaining camouflage paint becoming a distant memory, here is a hexagonal type 22 nestled next to the station bridge over a mill leet on the O gauge Butterwick layout. Pillboxes were employed wherever strategic protection was required for rivers, hill ridges and coastal margins. Great examples in combination survive at Cuckmere Haven in Sussex, Sidmouth river valley in Devon and alongside the GWR main line and canals from Reading to Swindon. (*Author*)

Perched up high above the western main line, tunnel portals on Martin Reynolds' OO Witham layout are being used as a tourist vantage point. This type 28 shellproof pillbox with six-pounder field gun embrasures has a great field of view and strategic reach (Hornby R8787 model). The pillbox study group (http://www.pillbox-study-group.org.uk/) lists all the types built by the War Department in 1940. (*Author*)

Here is a pillbox portrayed as newly built on the Shagbats O gauge layout by Mark Bamford of Sleaford MRC. Based on a seaplane base at Shoreham-by-Sea in 1943, this type 22 defends the rail access to the harbour. (*Author*)

A splash of Victorian military colour. As our Euston layout was based in 1875 for the visit of the Sultan of Zanzibar to London, we decided to have a red-carpet reception to catch the eye. Here are Arthur Wellesley, the second Duke of Wellington and Lord Lieutenant of Middlesex, with Lady Elizabeth. They are awaiting the arrival of the Sultan alongside the Coldstream Guards band. The white-metal trooper figures are from Langley Models and the pewter ladies and gentlemen from Andrew C. Stadden, figure sculptor. (*Author*)

RAF Kidbrook in OO showing what detail can be brought into a small layout with a military theme. The dog patrol where the wire is being replaced is a nice touch. (*Author*)

Mini project: Andover Road fuel depot in OO

Andover Road is based on the RAF fuel storage depot of Micheldever. This is a DCC bullhead track layout being built by the author with Hampshire third-rail electrics at the rear and disposal/tanker sidings to the front. As with the prototype the fuel storage tanks are situated in a rail accessed quarry (used to provide rock for the construction of Southampton docks) in the chalk downs and are covered by a substantial concrete roof designed to deflect bombing. It gives the opportunity to have a secure military area, convert standard road tankers for military use, make some Airfix kits, construct Ratio Nissen hut kits, etc. To keep motion there is a frequent electric passenger service in the background.

The images show some of the processes being used. Locomotives and rolling stock are always tested for clearance and position as the build progresses. Filament 3D printing is being used to produce the pipework and other bespoke requirements as a method of mass production. Previously a modeller would have created a master for a silicone mould and then cast resin examples. Foamboard and a surfacing of coarse sandpaper represents the concrete facing. Project completion will involve ballasting and a third rail for the arrival siding.

5
Fore- and Background

Above: BR (SR) Caterham station in Surrey basking in sunshine 20 June 1964. The running line to London Bridge Station is hidden somewhere behind, but this does not matter since there is so much to entertain the eye in the busy coal yard. To the rear the lush green trees on the ridge frame the image without detracting. To have a model train playing hide and seek behind features as it runs along your layout and having a sympathetic backscene can give more visual entertainment than a front and centre example. (*Online Transport Archive – Meredith-558-9*)

Right: Let us start with perhaps an aspirational backdrop given the space constraints on personal layouts. Here is the mountain range and thunderstorm depiction on the Coachella Valley Model Railroaders' HO layout situated in a large hangar. LEDs within the cloud produce lightning effects. (*Martin Reynolds MDMRC*)

Above and below: The Spirit of Swindon in N gauge, operated by the Little Layout Company, was test run in our Club room for the 2022 Stamford exhibition. The railway has great presence in both fore- and background detailing with the mass of Swindon station in between. The rear photo-realistic cloudscape is a challenge to deploy. It is a rolled seamless banner which is then Velcro-hooked into place onto vertical battens. For the individual with a fixed layout this is a potential game changer for the degree of realism and context that could be portrayed. (*Author*)

Below: The Sleaford MRC OO Bantry layout uses a clever 'breaking the fourth wall' technique. This fuel siding sticks out of the normal baseboard as a triangle to the public viewing side. This allows a longer perspective view from another angle to the norm, giving both an illusion of size and allowing detail to be under the nose of the viewing public. (*Author*)

FORE- AND BACKGROUND • 33

Sky backscene on a budget and to your own design. Here a south Lincolnshire sky – the Fens give a vast skyscape to view and catch the imagination. Choose an image to use as a guide for shading and colour matching your needs. (*Author*)

Above: The scratch artist. Basic acrylic colours from a hobby shop: cobalt blue, ultramarine blue, titanium white and Payne's grey. Watercolour paper of a heavier grade is used to avoid water distortions. (*Author*)

Above: The clouds are painted using a mid-broad brush starting with white areas and blending in a darkening palette. Grey is mixed with blue as the reflective and shadowed nature of a cloud picks up the surrounding sky hues. (*Author*)

Above: A cheap artist's brush set was used with a number of sizes and shapes. The finer brushes are utilised elsewhere with enamels, etc. for detailing models. A wash is blended onto the paper, giving vertical grading of sky from horizon up into deeper hues. Being shown is a first ever attempt – practice makes perfect. (*Author*)

Right: In situ behind the coal staithes of Butterwick. When the paper is stuck and flattened a spray of clear indoor acrylic varnish is used to physically fix the acrylic paint and prevent any damp damage. (*Author*)

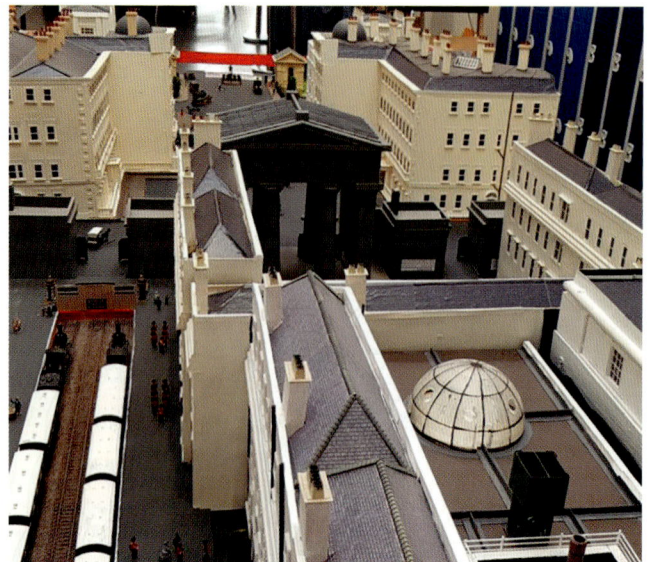

Left: Utilising a roofscape to tell a story. Here Euston Station 1875 in OO shows off the roof of the eastern booking hall with a glass dome made from a fish feeder globe. In theory this nested area could have been ignored and blacked out. However, adding a lightwell, water tanks, back windows of the Great Hall and enhancing the rearward view of the Euston Arch adds to the workaday station hidden behind grandeur. (*Author*)

Below: Phoenix Junction in HO by Mike Wyldbore plays a trick by blocking the view with a large foreground factory, forcing the viewer to look to either side and reveal new vistas of the freight yard behind. To the rear the eye is drawn to a candid view on the roof, with some incongruous light relief. (*Author*)

Seen from the rear before the backscene is erected. Spirit of Swindon at the rear has an extensive running shed, water tower and coal drop. Detail is present even where the public would not look. (*Author*)

Unserstadt by Brian Sibley in N gauge. Alpine displays with well painted backscenes pull the eye, showing a dominance over the railway. The shunting locomotive is dwarfed. (*Author*)

6
Water Features: Seas, Rivers and Puddles

Fishguard harbour with Brush 47 509 on 28 May 1975. The passengers will enjoy a smooth journey in the new Mk3 coaching stock. The *Duke of Lancaster* of 1955 vintage is alongside. Originally built as a passenger and short cruise ship she was rebuilt as a car ferry with room for 105 cars and 1,200 passengers. The service by this ship to Rosslare was short lived, being sold up in November 1978. The ship still exists, beached and abandoned on the River Dee. If you have concerns at portraying expanses of water in model form, consider viewing from rail side like this with a cutaway ship in the background. (*Online Transport Archive - WIckens 2794*)

Portraying Water on a Layout

If you look at the old war films of the 1950s where a sea battle takes place, one thing can be determined – the models are great, but water does not scale well. Aside from the mixing of water with the electrical element of a layout, all it seems to do is sit there and get dusty. Trying a real waterfall or water wheel on a model can be problematic. When it splashes, droplets are full size, so modellers have to look elsewhere. There are several options to choose from, including clear resins, gels, varnishes, Perspex and paint surfaces. Each have their personal adherents when executing a project and all can present a realistic finish with the correct treatment.

Left: Sandy Bay: North Yorkshire coastal railway set in the 1950s by Kevin and Maggie Smith. At just 2 metres in length, it cameos the view from offshore on a wonderfully calm day. The simplicity of the single-track wooden bridge is offset by the beachfront station and sidings, allowing for operational interest. (*Author*)

Below: The wooden railway pier on the Sleaford MRC OO Bantry layout shows an accurate portrayal of the prototype of the Cork and Bandon Railway (later the Great Southern Railway), on a plywood base with the angled sea wall coming down into the water. The shallow mud area with weed-covered rocks was built up and lower areas of wood were painted to represent green algae. Then a thin layer of resin was poured and agitated to give wave ripples during the curing of the surface. (*Author*)

Below left and below right: Dunstan Harbour with a storm at sea. This OO finescale layout by Alan Blackburn and Norman Cook shows a rough seascape on the Northumberland coast. The sea itself is sculpted and painted rather than poured to illustrate the rollers and whitecaps. (*Author*)

Mini Project: Butterwick Mill Leet in O Gauge

The middle board of Butterwick station was to have a millstream passing under the platform from rear of the board. Normal clear resin tended to dry with a misty finish, so a modeller-specific crystal-clear two-pack resin was purchased.

Technique was refined by some trial and error testing such as ensuring that if expanded polystyrene was present it would not dissolve, and seeing how turgid the mix would be. Also, whether multi-layer pours would work. This was so that leaves, fish and foam could be put at different depths.

Firstly, harmless experimentation with materials should be encouraged. See what works for you and what fails before you apply it to your layout.

Secondly, think outside the box. In this project below a serendipitous action led to a new way to create a basis for the water detail. Note things down or keep the material resulting until you need it.

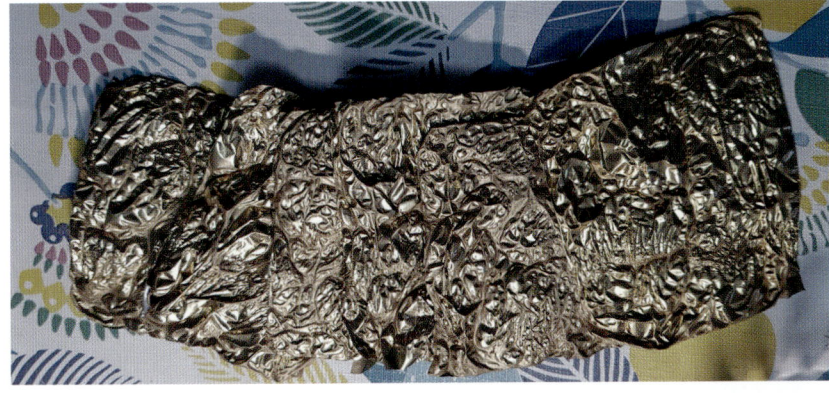

Serendipity: my wife placed a metallic crackly plastic wrapper from some sweets on the stove top of our woodburner for a moment (recreated for camera here), which could have led to a burning-hot sticky blob of plastic.

Right: This led to an 'Oh you know what that looks like!' moment and the heat-crinkled wrapper was duly popped away in a box for later use.

Below: The Butterwick station board had the concept of a millstream or leet coming down a waterfall and flowing out under the tracks and platform.

Having tested the capillary flow of resin prior to setting, and learning the hard way (it sets quite slowly and even a pinhole will allow quite a bit to exit your desired area), the area that would have the water effect was tanked using some old adhesive labels.

Embossed stone effect plasticard was used for the inner walls and undercoated. Then a bead of glue was used to ensure the area would not leak the resin when poured.

As you can see, the accidental creation of a plastic, textured base for water was repeated a few times and this was glued into position to represent channels and rivulets. Resin pours were performed over several days, building up layers after previous material had cured. In this case Solid Water from Deluxe Materials was used.

Acrylic paints were used to delimit different water flow areas: green for algae beside the waterfall where a turgid return flow would exist; brown for the mud channel sides; beige and remaining gold of the wrapper as the faster channel flow with silt and sand.

WATER FEATURES: SEAS, RIVERS AND PUDDLES • **39**

Scenic fibres from Deluxe Materials are a silky Santa beard-like material for waterfalls. The fibres can be used equally in a waterfall, rapids or sea-foam scenario. Here you can see the fibres being adhered into place with clear glue over the first layer of resin.

As layers of resin were added, white paint was used to add underwater foam from the waterfall. Also, different depths of weed growth, small fish and water-lily leaves.

The railings were 3D printed from a common licence design of a seaside pier. Many different artefacts exist either in a spares box or that can be printed and their use out of context can bring some unique results.

Finally, weathering of wall blocks and teasing in of ferns and other growth. Ferns and water lilies were cut out from green gloss enamel painted yoghurt pot lids. The thickness and texture of the foil tops enables leaves to be bent to position. The railings were aged with some rust powder, but also green algae wash on the sides facing the water.

A canal running from rear to front of the baseboard, emerging from under the running lines and into a brick-sided cutting on the Club Canons Cross layout. As the water surface visibility is minimal due to width and the presence of narrowboats the water is a murky brown painted layer followed by a varnish topping. (*Author*)

Portraying a canal can be a very useful device. You can have tunnels, bridges and narrowboats to add extra interest to a scene. Due to the narrow nature of a canal and towpath you get a lot in a little. This was a mini baseboard extension project specific to the canal and allowed a favourite pub to be included. (*Author*)

A culvert at the back of houses on N gauge The Priory. A simple bright gloss blue to represent cotton waste polluted water contrasts well with the static grass being scattered. (*Author*)

Above left: Coachella Valley MRC incorporate a busy harbour as variety in their extensive layout. Here the resin has been touched in with paint to accentuate a wave and surf line as the water shallows. (*Martin Reynolds MDMRC*)

Above right: It was mentioned that real water does not scale well. The alternative is to make the railway bigger and then it will work correctly. This example is a garden-scale layout at the Living Desert railway display in Palm Springs, CA. Here a full flowing river with rapids acts as the perfect foil to the railway. (*Martin Reynolds MDMRC*)

Right: Two views of Zweiberg by Brian and Ann Silbey, an N gauge river valley line. This has the visual trick of showing part of what is intimated to be a much wider river at the front. This is enforced by the river traffic barge and boat passing the town below the level of the running line. Water here is several layers of varnish with the wake of the boats added in plaster. (*Author*)

Nearly a puddle. O gauge Butterwick's water tower has lost its cover and is open to wildlife. When originally pouring clear resin, it was seen as being too transparent and thus not showing well. Just prior to setting it was dragged and a couple of waterfowl put into place to show the third dimension. (*Author*)

7
Vignettes

The Riddlesdown Viaduct precariously crossing an active quarry on the branch line to East Grinstead and Uckfield. A three-car Class 205 DEMU 'Thumper' unit in stock green navigates the plunging depths, surrounded by conveyors and work sheds on 5 March 1967. At the time of branch building, the quarry had already excavated part of the chalk ridge which originally loomed close to the nearside at camera height. Such a scenario on a model railway incorporates frequent railway services, vertical interest and industry all in one tight area and could be created as an isolated cameo model first, then expanded. (*Online Transport Archive - Meredith 642-3*)

The Art of the Vignette

A vignette can be described as a small illustration or photograph which fades into its background without a definite border.

If you are new to creating scenic displays or are attempting a certain scenic variety for the first time, it is good to practise. If your first cut of creating a small scene is successful, then you can incorporate it into your larger model. Many of the most successful club model railways are made up of a series of sub-scenes of a common look and feel. These are then stitched together to make a cohesive whole.

One benefit of working in this manner is that detailing can be performed in comfort, as opposed to leaning over a railway layout in the hot or cold at a strange angle. It shows in both the level of detail and the quality achieved that the modeller was not rushed, with all the access angles required, including from the rear, which perhaps a wall or backscene would otherwise have blocked.

Priory Farm: A Semi-derelict Farm Project in N

In book two of this series, prototype images were shown of a semi-abandoned Lincolnshire farm. This is being used as the inspiration for a lineside farm on the hill on The Priory N gauge layout. In N a lot of effort and detail can be placed on a piece of hardboard just 100mm x 250mm. This board once backed some brass detail frets and is being recycled. When the model is completed it will be blended into the hillside and field details as a 'step' in the hillside where excavations were made for buildings and a semi-flat platform was added.

Above left: The buildings to be used. Two resin printed pig sties and a filament printed part of The Priory ruin. Add to this a stable block card kit to add a splash of colour to the still operating area of the farmyard. So often you see this from a train, an area of investment and a remaining building slowly drifting to oblivion. A white metal lorry and horses have been purchased from P&D Marsh models as well to populate this area. The rest of the farm area is a dumping ground for redundant machines and will be overgrown. Cobbles and several other materials for walling come from the scrap box, leftovers from previous kits.

Above right: **Step 1**. This is a white metal kit and has a lead content. Wash hands after handing and safely dispose of shavings or filings. It is normally soft enough to cut and trim with a steel craft knife blade. Be aware that white metal can be either pliable or brittle according to the alloy ratios and ageing.

Above: Looking rather like a biological dissection. Pins or cocktail sticks stuck into polystyrene, cork or foamboard will support joints while two-part resin adhesive is curing.

Right: The body comes together and needs to be painted as an entity since the cab of the ERF lorry is separate. Likewise, once the wheels are in place the chassis can be painted. With model vehicles you end up thinking like the production line manufacturer. AEC, Leyland, etc. would often ship a bare chassis to a chosen body builder. So, you prepare your model in the same stages.

44 • MODEL BUILDING AND SUPER DETAILING

Aided by magnifying glasses with an LED light, the detail painting and construction is completed. A 2mm scale ERF cattle truck ready for the road at just 35mm long.

Moving key elements around for best fit once the Metcalfe stables building is completed. The space on The Priory layout is pre-allocated and this section will be dropped into place as a visual focus and scenery blended. For 2mm scale, painted figures from Woodland Scenics are being used. Sometimes less is more and a favourite building or figure will be omitted.

Decisions made. The divide between the sleek modern stables and older abandoned buildings begins to be accentuated. Admittedly at is point it does look like the planning for a disaster movie has taken place.

As noted, a useful facet of modelling vignettes as opposed to directly on a layout is that you can view from all sides in comfort. The primary uphill view is from this frontage. Once colour was introduced, the ruined pigsty looked far better as the first element to hit the eye. Inner walls were limewashed to highlight.

Above left: Static grass was applied, N scale trees are from the economy range of Penduke Models and understorey seafoam is from Woodland Scenics. All contributing to that feeling of the encroaching wilderness in this corner of the farmyard. The detailed rectangle will be blended into a road frontage and field system in the hill phase of The Priory's build. (*Author*)

Above right: Finally, an impossible image since the hillside to the rear will eventually obscure this camera angle. This is the long view as if standing in the grounds of the Cluniac Priory looking over the railway and distant mill town of Monks Beckbridge. The planning and build of this lower part of The Priory was covered in our second volume, *Constructing Buildings for Model Railways*. Young cat Holly shows that nothing is sacrosanct by chewing trees, but she is a good demonstrator of scale and perspective. (*Author*)

The Whole Layout is a Vignette

In a book of this size and broad subject matter we have to be selective as to what we choose to show. There are factories, mini wharfs, loading facilities, repair shops, etc. as elements for wonderful layouts. Taking things to an extreme in the space-saving scene with maximum impact is the turntable and roundhouse complex. Like many of the best scenic features they imply that there is much more out there, just unshown. The following two are great examples of this genre and both have 81A Old Oak Common as the basis for the model.

Above: The magnetic DCC trace board to allow the operator to keep tabs on which locomotive occupies which space. (*Author*)

Left: Seven Ash in OO by the Somerset Railway Modellers. Old Oak Common had turntables opened to the air in the diesel age, remodelling from 1964, with one surviving. (*Author*)

Left and below: Old Elm Park in O gauge by Mark Pollard of the Marlow, Maidenhead & District Model Railway Club captures the gloom of the steam age roundhouse. Built around a Kitwood Models 65-foot turntable model given as a Christmas present. The facility and environs grew to house locomotives that Mark built. (*Author*)

It didn't stop there. Due to his travel agency business going dormant during the Covid lockdown years, Mark expanded. What was a short fiddle yard became fully fledged locomotive disposal sidings and coaling stage based on Didcot by punching out through the wall of the shed into the light. The trick is that the roundhouse can be displayed on its own, with simple run out either side, or with the full extension and an exit out of the other side. Adaptability in your build means you can maintain your own interest and tailor your running experience to the space available at any point in time regardless of the scale. (*Author*)

A London Transport Scene

It has been many years since the core of the Club third-rail electrics layout Canons Cross was constructed. As a result there needed to be a practice piece built before a new layout – Andover Road, with Hampshire electrics and fuel depot – was developed. A small scene was decided upon based on a 'what if' of the Piccadilly line ending at Hounslow Heath. If you have the patience and space to create a practice piece the reward is a more polished larger project at the end of it. You can experiment with a variety of techniques, finding out what is best for you and the prototype being followed.

Preston Road station on the Metropolitan Line in August 1989, a bit down at heel but feeling corporate in suburbia.

London Transport (LT) has its own corporate look and feel whether on the cut and cover or the deep tube lines. This is due to careful planning after the amalgamation of independent companies and the formation of LT. The red District R stock at Northfields in May 1977 plus the poster for unattended packages with the roundel are evocative.

Baker Street Stanmore-bound Bakerloo service in June 1977 before the advent of the Jubilee Line. Source photos like these were the inspiration for Hounslow Heath, a terminus that never was. (*All images GWRPG/D Mulquin*)

A Sudbury Town box architecture based on the work of architect Charles Holden in the 1930s would be contemporary with this terminus. A long road frontage, stair access below and a large ticket hall window with a red bus outside would contribute to the recognisable atmosphere desired. The window was a rescaled 3D print of a much smaller French window used on the balconies of Georgian villas on the Euston layout. A set of unpowered EFE tube carriages had been languishing in store for a while. The size of the scenic box constructed was based on two car lengths offset and with Airfix platforms from an abandoned project. At the travel end, the lines are abutted by a mirror, making trains appear longer. To hide the mirror join, an overhead station building and access road was planned for the eastern end. A 'lazy Susan' TV turntable was placed under the scene to allow good access from a tabletop.

The trackwork being offered up plus the upper station approach being positioned. This has the station frontage and cambered roadway. It is made of scrap wood, card and 3D prints from various sources such as a seaside pier for ironwork.

The track was glued in place with PVA wood glue. Ballast was spread with a brush, then sprayed with water plus washing up liquid to break the surface tension. Finally, a 50% dilution of PVA and water was dropped into place to bind the aggregate.

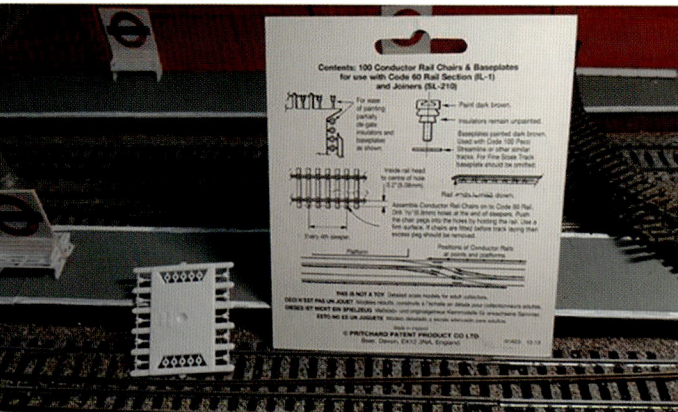

Above left, above right and left: Getting third and fourth rails into place. A ceramic insulator support chair is on every fourth sleeper, as can be seen in image of Covent Garden station. The rails were made using Peco rail chairs and redundant N gauge flexitrack that had been lifted from an older layout. The result is very pleasing and now requires grease, rust and iron dust staining added to the track bed and rails themselves. This is where learning on a small example is useful, for example insulators should be offset and not on the same sleeper for each power line.

More detail being added to the scene. The concrete fencing is from an LCUT Creative kit of coal office and staithes suitably converted. Despite LT being a somewhat shipshape urban railway, the weeds always get in somewhere.

Urban rail networks are advertising-rich – the aim is to anchor into the early 1980s. Note also the Hornby Dublo control room signal box rescue project from the second book in this series coming into use. Having seen the Bantry board idea, this box sticks out over the 'fourth wall' in similar fashion.

A passenger-eye view on completion. If you want easy lighting, technology has moved on to help. Look for the strings of small battery-powered LED bulbs. They can fit easily into a scene to light it all.

The streetside view. The road was made from card supported centrally and cambered. (see the website http://www.constructioncivilengineering.com/road-cambers-and-its-types.html for the key definitions). Poster hoardings like those adorning some familiar commuter stations contain period images and, with trees, soften the backscene transition.

Light Industry

Through urban zoning, historical positioning and spare land becoming available after railway closure, goods yards, industrial sidings, light manufacturing and commercial units all become possible.

There are many models available, to use in original form or modernised to represent a change that may have occurred through time. It also opens the door to incorporating a rotation of scenic additions for those who enjoy making models. One day it's a goods yard on your layout and the next, with a swap-around, it becomes a vehicle repair depot.

On the Club Canons Cross layout extension we had a section of land kept blank near the church, simply grassed and open. It enabled us to swap over for a model village, a car showrooms or a seasonal fairground. It keeps the model fresh, and the variation of a scenic nature combined with changes in rolling stock can impart the feel of a new layout.

Where possible, think ahead. Don't always stick your buildings down. The light industrial and commercial areas of the railway, where there is a sharp demarcation between sites, lends itself to swapping buildings out. It is harder to achieve in a suburban setting due to gardens and street furniture.

The area around the coal staithes on the O gauge Butterwick layout was seen as a dormant corner which did not pull the eye. Models were sourced to bring in light manufacturing and repair, as would appear in a rural economy. Although 1:43 is desirable, some models at 1:48 will fit nicely into limited space without looking odd.

Beyond the buffers a closed workshop existed. Perhaps enigmatic, but the corner felt dead. Any vehicle placed here that was not a coal lorry looked wrong.

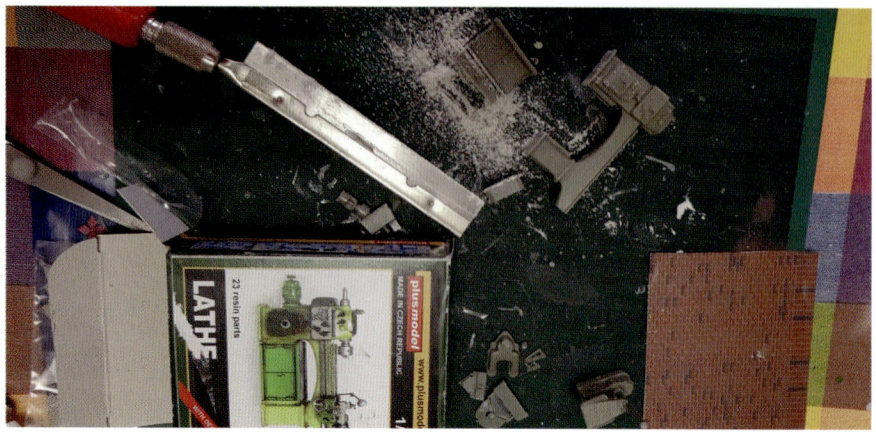

Interior details are good in any open-door scenario on a model. They give the viewer a feeling of depth and can express a story behind the installation. Here a resin model lathe is being trimmed and constructed with micro saw, file and two-part resin adhesive.

A 9V battery-driven welder's torch LED circuit with bright flickering white followed by fading red was sourced and positioned deep inside the workshop. Brickwork flooring led into the viewable area with the lathe and pillar drill models.

The interior equipment painted and ready for positioning. Younger viewers have surprisingly acute eyesight, they pick up on the tiny details that get added to a layout.

In action: gas bottles on a trolley, tin foil at rear reflects equipment and light, giving a greater feeling of depth. Tractors, rusty cars and bulldozers can now be positioned as desired for a different display each day to invigorate the corner that was once dead.

When film production and our participation in the short film *Dream Big* had wrapped, the Canons Cross layout returned to the Club room. There was a gap where a student contribution had been returned. It was decided that this could be the 'Fairfield' of the layout, where different drop-in elements could be used to change the look and feel of this static part of the townscape between exhibitions. For the show following vandalism and Covid lockdowns our main sponsor was a car dealership. Their site was surveyed for modelling by Club member David Hildred. A combination of site photos and the use of satellite images gave size and position.

The completed car dealership dropped into the 'Mirandaville' section of the Canons Cross layout. Finished with a nice array of the Ford product from the Oxford Diecast range. (*Author*)

Above: Models can also convey drama. Here a desert scene by Alan Hancock has explorers, derailment and peril all rolled into one, with a quirky Backwoods Miniatures On7 locomotive. (*Author*)

Left: Club OO layout Amberdale incorporates an allotment gardens display which was built as a separate entity for another project. Handmade paper leaves and micro netting add to the realism. Lineside allotments, whether for staff or local residents, are often forgotten, as are the less picturesque sewage treatment works and suchlike. Since this layout was designed for a magazine project, time was of the essence. This detailed section was selected and let into the baseboard, looking as if it was part of the core design. (*Author*)

Heavy Industry

The railway historically serviced much of the heavy industrial requirements for both raw materials and redistribution of finished products. For many areas this, rather than passenger traffic, was the reason for the railway.

You can decide on a specific industrial sector to base an entire layout upon or intimate with backscenes and other extras that there is industry in the area of your model.

Unlike light industry and commercial, where you can choose to adapt and swap, once you've decided on installing sidings and factories you have them resident for keeps on your layout due to the depth of infrastructure.

Right: A 1970s power station in N gauge by Ken White. With two large cooling towers, conveyor belts, modern office blocks and a rear fiddle yard, this layout shows what can be achieved when based on a sole industrial sector. Inbound 'merry-go-round' block trains either of VGAs for coal or tankers for heavy oil sweep into the main viewing area to vary the interest to the operator and viewer. (*Author*)

Below: The Westgate industrial area at the San Diego Railroad Museum. If you have the room to dedicate to an industrial area, the use of uncluttered white space is an asset. (*Martin Reynolds MDMRC*)

The Port of San Diego with oil storage facility at San Diego MRC Balboa Park, CA. As can be evidenced, if you have plenty of space you can spread out and occupy it. In the UK most modellers operate under domestic constraints so alternative set-ups need to be considered. (*Martin Reynolds MDMRC*)

A simple but effective fuel depot on the HO scale layout Padden Flats by Mike Ford, depicting Wyoming in the 1960s. (*Author*)

The opposite end of the scale is the refinery and electrical substation on the Coachella Valley MRC HO layout. (*Martin Reynolds MDMRC*)

8
Railways Within a Scene

A gentle scene of a rural light railway, in this case at Vrads Sande, 1985, the terminus of a short preserved line in central Denmark. This sandy area lacks ballast and has few intrusive weeds, just seeming to coexist and sink gently into the surroundings. At busy exhibitions some of the most noted layouts are those that could be framed and put on the wall. The railway does not dominate, they present a suggestion of the iron road within a much greater landscape, and thus become artworks in themselves. (*Author*)

The Scenic Journey

With three books produced giving concepts and directions on how to build, it is perhaps time for a simple gallery showing model landscapes with the railway blended within. Let us, dear reader, stimulate the inner Constable or Turner. It takes a certain confidence to reduce the railway to a subset of the whole scene, limiting the 'play' aspect of the model into something quieter and bucolic. That may be an end in itself to bring relaxation. Sometimes simply listening to a train going round and round without intervention can be hypnotic and relaxing. At other times immersing oneself in a landscape can achieve this, both through the modelling phase and the post-construction enjoyment.

The aim is often to impart the feeling of something as an interpretation rather than being photo-realistic. At times these are

second layouts or showpieces. If space and circumstances allow, the main layout has the complexity and wow value. The alternate layout allows the artist to emerge. The location, weather and climate can vary allowing the common midsummer in busy 'Blissville Junction' to become something that is unique, eyecatching and artistic.

Two views of a snow-covered end of the line scene on the OO Southern Railway at Humphrey Road Sidings by Norman and Meg Raven. It gives the opportunity for simple train siding operations to look special and the scene itself to stick in the mind long after a show. Attention to detail where snow melts around the chimney stacks, cold colours outside and warm interior lighting all add to the winter theme. (*Author*)

Let us take things to the opposite extreme. If your layout is big enough the railway gets lost in it! The crew at the world's second largest model railway museum at Greeley, Colorado, has a scale 20.5-mile long HO layout through the Rockies. Admittedly 300 locomotives and 2,500 freight cars are needed to service it. You too could try this if you have a spare 500 days and a desire to make 28,000 trees. (*Paul Ijs MDMRC*)

If you have enough outdoor space – and admittedly here in Palm Springs CA some decent desert conditions – you can consider making use of the garden terrain to blend in a substantial model railway. (*Martin Reynolds MDMRC*)

Above left and above right: Back to reality for the UK, here in N gauge is pueblo detail from member Alan Hancock's Gila Canyon 1957 layout. Rather than using 2mm scale to crowd in detail, there are spaces with eye catching New Mexico style pueblo settlements. Plus, a visiting tourist party while the railway meanders past. Little extras catch the eye, such as the abandoned vehicle in a dry gulch. (*Author*)

Above and below: Three views of the club Euston Station 1875 layout. Although this display will get a dedicated fourth book in this series for both model and site history, it is worthy of inclusion here. First progress from the New Road at Euston Square Gardens. Then travel down Euston Grove past the Joseph Bazalgette's pipe depot for the new sewage system and some imposing Regency villas, before marching further north between the sheer cliffs of Victoria and Euston hotels on Drummond Street and reaching the imposing Euston Arch. Only then can you contemplate that there is actually a model including track and rolling stock carrying on for the same distance beyond. The eagle-eyed here will spot the 3D-printed French windows that were enlarged for the tube station Hounslow Heath earlier in this book. There is also a soundscape for use with micro-USB speakers in the station. (*Author*)

Six views of Neuberg in HOm. Twin Swiss layouts by Jim Finlayson, showcasing the same terrain but in 1913 and 2013, thus cleverly portraying 100 years of change. (*Author*)

Mini Project: North Cape, Kimberley, desert river

Book one of this series covered the construction of modular baseboards with adaptive lighting. LED strips with a dimmable front white light and multi-colour rear uplighting were utilised. The aim was to bring a showcase for a South African scene allowing for hot desert daytime fading through to late evening sunset.

Several geographical scenes were to act as focal points, this being a scene loosely based upon that of the Modder and Riet Rivers in North Cape. The area has a mix of arid plateau desert, narrow riverside wetlands and industrial sidings.

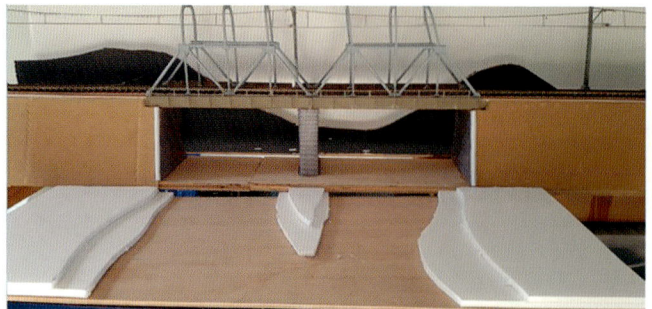 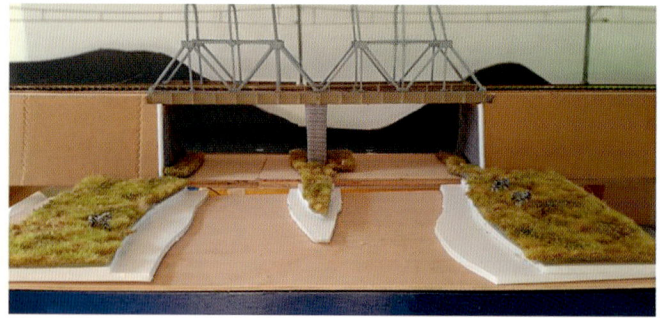

Above Left: A 3D filament printed river bridge was scaled to HO, with Piko catenary for the overhead power pantograph pickup strung through the scene. Foamboard was used to build the river terraces. *Above right*: The terraces were populated with fine ballast, painted cat litter chunks for rocks, self-adhesive grass tufts and autumn mix 6mm static grass as the main substrate, some applied roughly by hand.

Sunrise brings a fuel train. The riverbed was made from domestic patching plaster plasticised with added PVA. Clean clay cat litter was used to add the rocks in the river. Once dry, a bead of clear silicone was added to the front to prevent the resin being used for water running away. It was then painted with acrylics of several colours, merged on the riverbed while still wet.

Static grass has been a game-changer in terms of the realism that can be achieved. Here a small scenic detailer is being used. The crocodile clip grips a steel nail which completes the electrical circuit when it touches diluted PVA brushed into the scene. A few vigorous shakes and the static affected grass filaments land upright. Here 4mm spring and 6mm wild meadow are mixed to make the riverside slightly more verdant.

Endgame. The Blue Train passes at twilight heading north to Kimberley, while tourists in a Land Rover catch the wildlife. Two-part clear resin water effect from Deluxe Materials was poured in several layers allowing some deeper white water to be painted in and long weeds to trail from rocks. The small Cape Mountain Zebra are N gauge horses from P&D Marsh carefully striped with paint on the end of a cocktail stick. Yellow flowering brandy bush (*Grewia flava*) is represented by World War Scenics daffodil tufts. The image is unadulterated with the sunset effect from rear RGB LEDs and front white light 75% dimmed.

9
Detailing Rolling Stock

The lovely detail of the diminutive broad gauge South Devon Railway locomotive *Tiny* on the platform at Newton Abbot on 18 July 1959. If you are new to the concept of detailing existing models, start small and simple. Build your skills up over time and tackle the larger projects once you feel confident. (*Online Transport Archive – Meredith-75-2*)

While there are detail upgrade kits for older rolling stock there does come a time when the question is asked: Why am I doing this? Looking at the detail improvements between the old Lima (at rear) and new Heljan Class 33 locomotives, even if DCC and lighting is discounted, the old would take a lot of skilled work to bring to modern levels. The answer often comes back – because I want to!

Mini Project: OO Victorian four- and six-wheel coaches

For a club or indeed an individual to provision a main line London terminus station with sufficient rolling stock for a set time era is a challenge. Recently, both Hornby and Hattons have come out with detailed generic coach ranges in a number of Victorian liveries. For the Club Euston station, the front of house tracks benefit from the detailed examples of ready to run, but to have enough for the whole station to be populated would pour too much resource into a single project, therefore an alternative was required. The Hornby starter set 'bug box' four-wheel coaches needed to have a simple cut and shut to make longer four-wheel and six-wheel rolling stock. Since they are regarded as toys they appear at shows as an inexpensive second-hand purchase. As these were deeper into the diorama, they could be hand-painted into LNWR livery and did not require spray painting, detail lining or transfers as they are out of human eye detail focus in the backdrop.

Right above: They have been around for many years as a part of kids' starter sets and share the underframe with the 20-ton brake van. For our purposes the existing livery paintwork or base plastic colour does not matter as it would be primed.

Right middle: With some scribing and the use of an Exacto micro bladed manual saw, the coaches were cut to size. The glazing strips include the roof detail, so they were reversed to add strength to the chassis overlap by a compartment's length of roof and window. You can see this on the red and blue example centre left. A brake van was also cut up to provide guard's viewing duckets. The roof rain strips were sanded off and replaced at full length with a square section plastic rod.

Right: Once stuck together with two-part epoxy resin and gaps filled with plastic granule filler they were sprayed with automotive grey primer. Finally, they were hand painted with Precision Paints colours. They appear slightly crude against the modern finish of the Hornby LNWR coaches but in the background they work well.

Below: Behind Improved Precedent 2-4-0 1673 Lucknow on Euston platform 6 is one of the basic detail coaches made from this method. The tank engine is a 'Smokey Joe' Neilson type loco with the cab cut off. It will soon be lined and lose the outside cylinders, to represent an LNWR Ramsbottom 'Four Foot' 0-4-0ST shunter. Minimal conversion to make the purist shudder, but it will just be visual infill for shows.

Adapting to Your Own Prototype

Many people have favourite items of rolling stock, whether due to the design, livery or another attachment to a specific class or just a single example. The challenge is often to make it relevant to the layout and justify its presence. Departmental stock can often do that, or a change in commercial use that occurred. Here are three vehicles that the author has in his collection that are good examples of this – every picture tells a story.

Above left and above right: The BR built Blue Spot fish van was a long-lived vehicle, but not in the original intended service. As fish traffic reduced many of these were seconded to parcels service. This example ADM8787 ended up in preservation as a generator van. The author, with his wife alongside, used this power for needle guns to clear the paint on the Southall sheds footbridge. This involved dangling over the GWR main line in a scaffolding cradle. This is an O gauge Parkside model liveried in parcels use with Fox Transfers for Butterwick's portrayals of the 1970s. (*GWRPG / Author*)

Above left and above right: Hornby R6970 ex LNER extra-long CCT van. The three sliding doors plus end loading made this a very useful preserved item when it came out of revenue service in the late 1970s. This is the prototype of this CCT that was measured by the Hornby designers. The vehicle starred in the film *Trainspotting*. In 1995 the shed at Southall was used as a film location for a night scene. The roof of the van had a graffiti tag portraying a giant half-opened fish can with ring-pull opening. This van is used on the Andover Road fuel depot layout and is due for some gentle weathering. (*GWRPG / Author*)

Above left and above right: Ex LMS and latterly seconded to electrification in the 1990s is 21-ton brake van M731211 dating from 1941. The author and his wife celebrated a wedding anniversary by buying/donating this vehicle before the cutter's torch reached it. The Slater's model in O gauge retains the stove – someone pinched the 1:1 original! The van is now safely resident at the Avon Valley Railway and the model runs on the author's garden line. (*GWRPG / Author*)

Weathering

In real life, rolling stock wears out through age, traffic type and finish applied. Some profile revenue stock would see a regular cycle of refurbishment but taking the four-wheel short wheelbase van or open truck in British Railways as an example, from pristine in 1950, it would be unlikely to see an interval repair or repaint, ending up as a piebald rust bucket by 1970. China clay wagons would be constantly coated with a white powder, coal wagons gravitate to black dust and rust, shunting locomotives would become a greasy, rusty seemingly unliveried mess. The railway modeller starts with a pristine example and then has to bite the bullet and decide how far to take things. The opportunity can be taken to create doppelgangers at the same time. Renumbering stock and diesels on different sides to make it appear that you have more stock and varying the style of weathering adds to this. Keep dust and cobwebs that build up – they add to the feeling of overuse.

Dapol O gauge class 08 duo shunt the yard here with a trestle wagon and a bolstered oversize Warwell load. Loads themselves vary between the pristine and the long stored/reused and need attention as well. The locos have been airbrushed with several layers of track and dirt colours. Working out airflows and oil deposit areas as well as paint fade can be tricky. Compare to the cleaner Derby lightweight DMU and Class 40 D213 *Andinia* behind.

The opposite side of this shunting duo, numbering and arrows of indecision locations are different as are the wheel motion colours. It is harder with a steam engine as often the smokebox has a number making the side-to-side identity split impossible. You just have to buy twice as many!

Mini Project: Detailing the Tri-ang R226

Produced over a number of years, the Tri-ang Southern Railway Maunsell bogie utility van was made in maroon, green and blue injection-moulded plastic. The prototype included one vehicle liveried for the Pullman service that carried Sir Winston Churchill's coffin. With six opening doors this model has stood the test of time in the quality of the moulding, especially in the upper body. Where it particularly lacks decent detail is in the bogie/wheels, underfloor and carriage gangway ends. Having purchased a green example that had lived in a nicotine-rich environment, it was decided that it would be upgraded by use of the Roxey Mouldings super detailing kit. In theory the finish of blue didn't happen as surviving vehicles went into the service fleet. However, memories of newspaper trains being passed in daytime sidings ready for the next night's service loom large, so that is the overall target. There are times when you just have to say, 'My layout, my rules'.

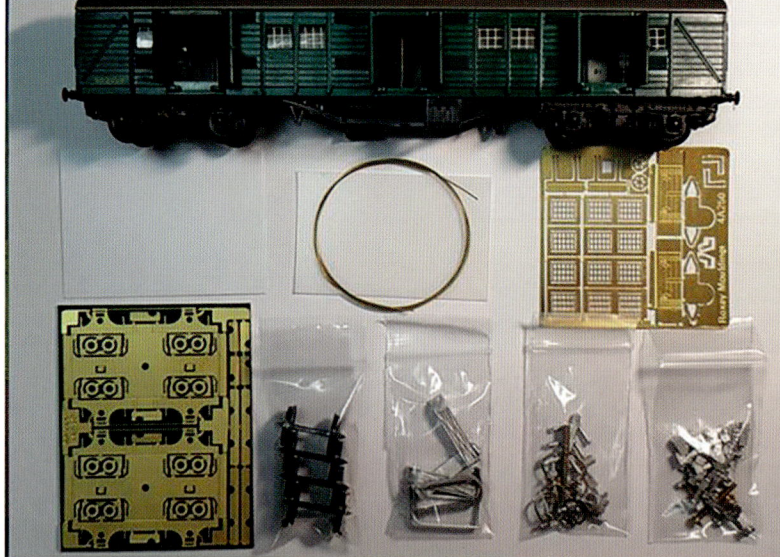

Above: The donor kit on the right and a rail blue plastic moulding example to the left. It had all doors working but that would change to fixed ones to make them more realistic. Couplings were missing. They would have the Kadee (www.kadee.com) variety replacing them. There is an option to change the roof profile as well as replace the torpedo vents. Certainly, the vents would go, but at the start of project the jury was out on the roof profile change.

Left: While doors open in loading mode is a rarity on model railways (Mainline did a sliding door LMS 12t van, Hornby similarly on the ferry van, but few manufacturers followed suit), they will be fixed and detailed. The physical detailing parts are good quality brass etchings and white metal castings originally produced in the early 1980s. You begin to understand your materials and their limitations. Brass will bend and needs cutting with shears. White metal looks like it will bend but unless very thin it will break. Plasticard cuts with a knife and can be snapped along a scored line. The kit comes with a detailed plan. With the vehicle turned over the bogies are to be drilled off the chassis and the central detail removed. This proved to be surprisingly brittle plastic, so just as well not needed.

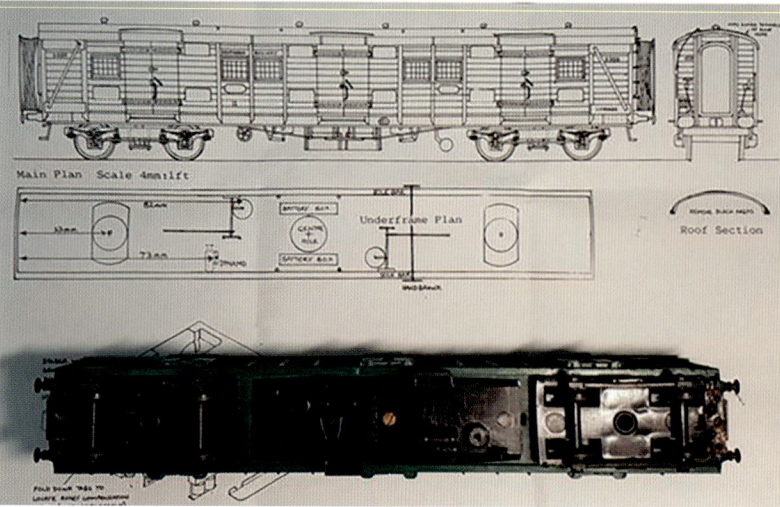

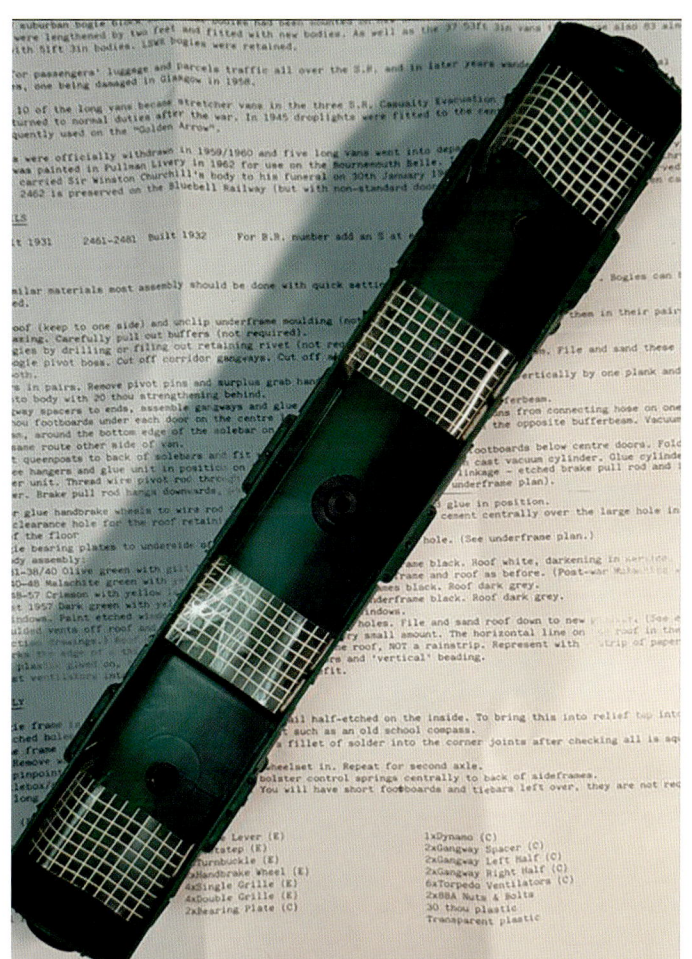

Roof removed from body to reveal the white grid plastic that forms the glazing. On removal the white paint began to spall off, showing the fragility of the material. This is replaced by finer grills from supplied brass etchings.

Disaster! When drilling out the bogie retaining rivet, the drill bound, the body twisted, all combined with an autonomic reaction to hold tighter. This being an older Tri-ang example, the plastic is brittle when cool. Lesson (learned far too late) is to keep the roof in place to retain rigidity.

To the rescue, the blue example. Side cut shears used to trim the top hinges off the doorways and a sharp blade for the lower ones. Doors have the lower handles removed and are permanently stuck together. Then a new plank was added to the door tops from plasticard.

The brass fret for the bogies was bent to shape with wide-blade tweezers and white metal accessories glued on with two-part resin. This glue is best when multiple material types are being used. The downside is that even the rapid versions act like grease until they set. Be prepared to use clips, weights, braces from any medium you find useful until the joint is solid. Here, when putting finescale wheels and brass cup bearings into the brasswork, one sprang out over the floor and under the fitted bookcase. So here is another useful modeller's tool: a handheld vacuum cleaner so that you can then go through the dust to find lost parts! Be pre-emptive and assemble bits that may spring apart in a large plastic bag.

At the build commencement the carriage got a gentle scrub with washing-up liquid and water. Now deep clean time. A cotton bud soaked in solvent to get into every nook and cranny. It is amazing how much dirt there is on what was a child's toy. The supplied glazing and grills will go in next. The roof inner support will need trimming to allow for the glazing, thus retaining a flush joint to the sides. While the rail blue paint from Precision Paints will be hand brushed on with a soft sable bristle brush, the weathering will be sprayed on after application of transfers. A note on paintbrush lifecycle. Have an array of modelling/artists' sizes and shapes and ensure you 'progress' them through their life. If you start off with a fine brush, move it through general thicker and darker paints until you often end up using it for applying adhesives or plaster texture finishing before binning. Don't be tempted to short cut and compromise.

The roof vents were replaced with supplied castings, but the actual roof profile was left alone as a task for one day when more experience and bravery exist. The transfers from Fox are refactored from their GUV examples to fit this imaginary repurposing of the vehicle. Interestingly, the example at the Bluebell Railway was condemned and purchased from the scrap line at Micheldever, which is the prototype fuel depot for Andover Road, seen here.

10
Building Your Own Rolling Stock

BR (SR) Class O1 locomotive 1390 in store at Ramsgate, 21 April 1951. This was a 1903 Wainwright domed boiler and new cab rebuild of the 1878 Stirling SER Class O. Although we are not covering locomotive building in this book the overall premise is that kit or scratch building is the option for your niche requirements. An impressive number of new ready-to-run products covering many vehicle types are being developed, infilling some of those troublesome gaps. Often, unbuilt kits or older donor items can be sourced cheaply and reworked. Using eBay or visiting second hand stands at shows can unveil bargains. (*Online Transport Archive – Meredith-177-3*)

Having fun with kits hones skills and gives confidence. Here are some variations on the diminutive Airfix (now Dapol) Pug saddle tank. The author and his older brother had a bit of a competition in the late 70s. Anyone for a Pug double Fairlie?

Building Model Wagons

This book is aimed at the entry-level railway modeller. There are specialist publications aimed at brass locomotive and carriage building, which requires advanced skills. Model wagons are the ideal start point for building up kit building skills. Here we make use of the wagon building experiences of Club President Colin Brown. He knows his prototype well, having spent a long career on the railways, ending up as senior manager in the wagon and tanker leasing business. On the modelling side he has helped write instructions for some O gauge kits on sale today, built well over 100 of them himself and contributed articles to *British Railway Modelling* magazine.

Above right: A more challenging kit from Slater's with the next degree of realism, intermediate rather than starter due to the small brass detailing parts.

Below: The Parkside LMS van donor kit for this build example. While demonstrated in O gauge the techniques and skills transfer well to OO. The kit is really a simple box with wheels with good detail injection moulding, decent wheels and working buffers.

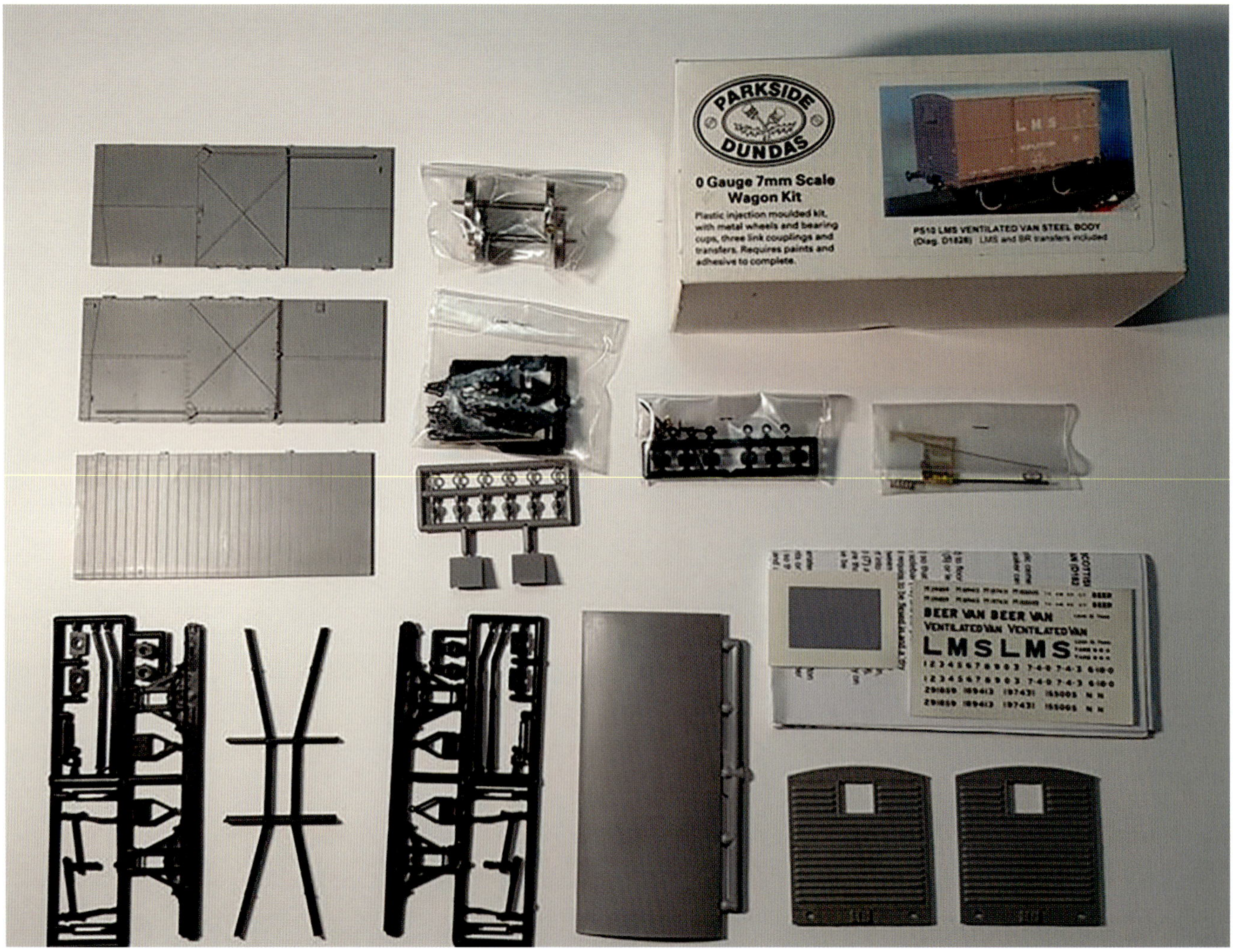

To start at the very beginning

Let us step back from the act of building a kit and cover some of the terminology and tools. Please do not be offended at this section if it feels like being 'taught to suck eggs' – there will be some who are true beginners. It can seem daunting to start building your own kits, the ready-to-run offerings are so good nowadays. It is also very frustrating to start a model wagon and end up with what feels to be an inferior result after much time and effort. When choosing your first target, keep it simple. Aim at injection-moulded plastic as a single build material in the kit. It can get difficult trying to choose which adhesives to use as well as understanding the differing handling characteristics of the materials used for super-detailing. Even if the resulting item is not exactly what you need, it is a learning curve where complexity comes later.

Having chosen your plastic kit your next requirement is a safe, flat, protected and well lit working area. Safe, in that it is well ventilated and that other family members do not get unwanted access to harmful sharps and liquids. Whether a shed workbench, kitchen worktop or dining table is chosen, you should have a clear area of approximately 30cm by 60cm. This means you can move items around and are less likely to damage them or yourself. Ideally you need a flat piece of board or ply to ensure the chassis and wheels are flat as you construct. Also treat yourself to a self-healing cutting mat or specialist glass board from a craft shop or model supplier.

Basic tools include a craft knife (rather like a surgeon's scalpel with various shaped blades), tweezers and needle/jeweller's files. Useful are very small diameter drill bits (0.5mm to 2mm) and a powered micro drill or manual twist chuck drill. Additionally, small side cutters, a set of different nosed small pliers, a steel scribe, crocodile clips and a fine toothed Exacto-type saw are good aids.

Get some padding, whether foam, rags or other material, so that if working with the model inverted no damage occurs. If you can find a small puffer such as that used to clean cameras or a redundant ear syringe bulb, it is ideal to use to blow away dust without introducing contaminants and moisture.

The type of glue/cement used depends on the requirement of the join to be made. A liquid poly brush-on type will work well on most joints (Revell Contacta, Mek Pak, etc.). In some cases, a gel type general glue (UHU, Bostick, etc.) will be useful to spot on or run a seam.

Read the instructions through thoroughly, annotate if you need to. Look at images of the prototype in books or online. Sometimes instructions can be a little sparse, misleading if there are variants involved and alternative options from the sprue. A sprue is the waste plastic moulding pipe that feeds into the detail for the kit. Often a series of fragile components exist attached to the sprue requiring careful removal with a micro saw or side clippers. Check the content of the kit, it is rare that anything is missing or broken, but now is the time to be sure.

Ends receiving a little treatment and using liquid poly glue to stick on the vents. The glue bottle top is always screwed back when not being used and kept on an old pant mixing saucer so spillage is not a disaster. The pointed blade of the knife proves very useful as it can also ream out holes. Fine tweezers and the puffer stand ready.

Commencing the build

With vans and wagons you start with the carcase cube and add the chassis and details on after (unless the instructions state otherwise). Where possible use either a fine-toothed saw or side cutters rather than a knife for any sprue removals. This way you avoid distortion and the danger of snapping finer components through twisting. Remove floor, roof and the four sides for inspection. It can be useful to determine the edges that will be adhered and key them slightly with a file or sandpaper – the glue will take better. Also check for blemishes or sprue joints that need removal and clean them up. Dry fit the box together to determine the fit is good. You can use a weak elastic band to keep the unglued box together if desired. Make sure you have the sides the correct way up, or with a wagon any opening ends at the correct end. When the fit is correct use a fine brush for the liquid glue to bead both surfaces. Then offer up the joint squarely. Normally the glue will have softened the plastic and joined inside one minute.

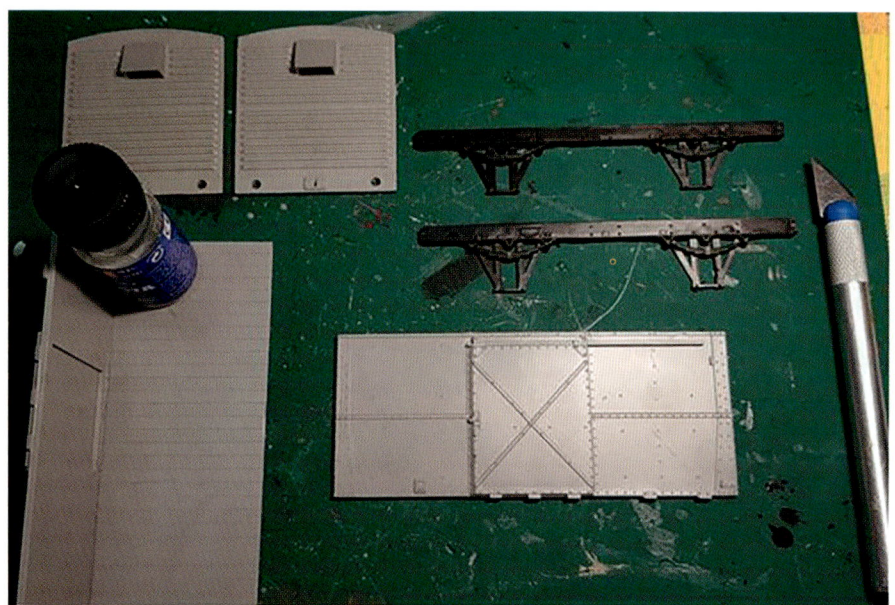

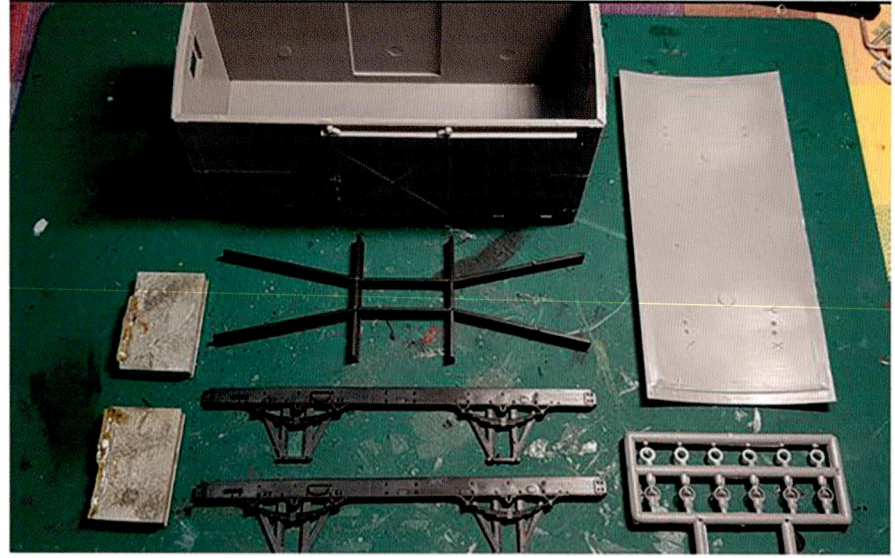

The first van side being stuck to the floor after roughing up with a file. The glue bottle is a 90° type so is being used to ensure the two parts are correctly right angled as the glue dries. Once the carcass is constructed and prior to sticking the roof on, weighting is being added inside. Here some scrap white metal, but fishing weights, plumbing/flashing lead offcuts are all good. The aim is low centre of gravity over the axles. Use a two-part epoxy resin glue as you do not want anything to come loose in a sealed model. Also note to the right there are torpedo roof vents provided, but this kit does not use them. It means that the markings on the inner roof for drilling need to be ignored for this kit. They go to the scrap box and may appear in OO on a model warehouse roof.

1. Axle box, brass axle bearing and back plate.
2. The back plate and the W iron on the chassis are filed to ensure smooth vertical travel.
3. Back plate and bearing positioned ready for gluing.
4. A blob of gel glue is placed on the bearing head and the aim is to glue the brass bearing inside the axle box without polluting the W iron. Clean vertical travel compensates the axle for rough track and points.

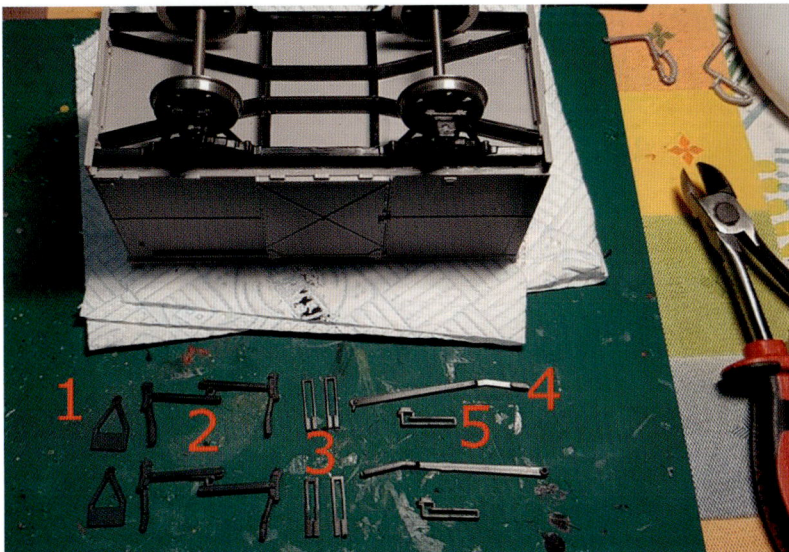

The side clippers have been used to remove finer sprue details.

1. The brake central V hanger.
2. Brake arm push rods and brake shoes. To glue in line with the wheel rail faces.
3. Four safety straps which slot round #2 to catch broken brake arms.
4. Manual brake lever.
5. Toothed rack to retain brake lever. These have a habit of breaking when being positioned around the lever, so treat with care.

This model is having extra white metal vacuum brake connectors and cylinder added as some were continuous braked later in their lives.

Paint shop. Carriages and locomotives deserve consideration and you should practise hand spraying paint in a booth. Freight stock will often require some weathering, so hand painting can be the first step in the non-perfect finish. A toned undercoat, matt white for unbraked grey and matt brown for bauxite worked for the real railways and works well here. Then a Precision Paints topcoat goes on. Stir very well in order to thoroughly mix and then use a broader quality brush to get an even topcoat into place without brush marks. Transfers: waterslide are the easiest to use, Methfix or Pressfix give a better finish but are harder to utilise. Learn what the lettering represents and where possible refer to a multiple prototype photographs as rules are broken. For a generic BR example:

A. Freight maximum carriage weight in tons.
B. Prefixed number. P for ex private owner, M, E, S, W for originating Big Four builder company or B for BR.
C. XP, continuous braked stock > 10 foot wheelbase, Class A train capable (passenger, fast parcels, etc.).
D. Overall wheelbase limits the route and speed.
E. Tare weight of the empty wagon ton/cwt.
F. Any special branding, experimental, return to, etc.

Offered up to the layout but transfers not yet overcoated with satin varnish, with some weathered unbraked brethren behind. Weathering is a gentle art – the amount applied depends on the age of a vehicle, the type of route and load, whether it was ever washed, patch repaired or rebranded. There are some great YouTube and website explainers online so please take advantage of them. Freight stock is easier mentally to take from pristine to weathered. The impressive livery finishes of more expensive locomotives and carriages take some bravery to weather and you would be better off having practised on your cheaper freight vehicles first.

Remember to weight your wagons and compensate the axles, especially on longer wheelbase items (compensation is a manner of allowing some play in the axles to allow for track irregularities). Otherwise, derailments will happen in forward, but especially in shunting motion. The images here were taken by the author in Southall WR depot in the mid-1990s. Here a long Spie Entrains consist for Powertrack Heathrow electrification has been derailed adjacent to the footbridge when being yard shunted by main line liveried 37 798. The buffer lock rose vertically on the empty open wagons, but the couplings held until pressure was relieved. The Old Oak Common rerailing crew soon had the DPT generator van and OBA/ZDA wagons block packed and levered back on the tracks. Meanwhile a long line of mixer bogie flats with the concrete mixers for catenaries went off and got dumped in the yard. An expensive error with the whole rescue crew on overtime plus callout fees. The pointwork traversing the main line behind is where an aggregates train was hit by an HST in 1997. (*Author*)

Mini Project: Carflat in O Gauge

Time moves on and untouched model kits build up in the loft or shed to be attempted when time and skills allow. Then manufacturers come along with highly detailed ready-to-run examples, removing the original premise and rendering the investment redundant. That is how it felt when two Easybuild 7mm carriage kits emerged from the shed during the Covid virus lockdown along with collected 1:43 car models of Triumphs from the 1960s. The brainwave! An ideal combination to display them running on layouts as a factory train.

Open carriage wagons date back to the dawn of the railways. The Duke of Wellington did not like train travel so always rode inside his carriage while on a flatbed truck. British Railways built carflats in the 1960s based on redundant coach chassis for both the precursor of Motorail services and also for factory vehicle distribution of chassis, part-builds or finished items. Initially pre-nationalisation and latterly BR Mk1 bogie coaches became donors. These kits therefore became the ideal lower wagon parts. Then 3D printing and spare rails provided the less substantial upper works.

A quick online visit to the excellent Paul Bartlett photo archive on the internet to see some prototypes and the project started swiftly. The carflats retained their original bogies in the conversion and were awarded a 10-ton loading. While Boflats and Boplates are of a similar configuration they tend to have specialist wagon David and Lloyd type bogies. Mk1 conversion St Rollox 1968 diagram 1/068 lot 3679 was aimed at.

Right: Parts of the unopened kit awaiting assembly, not shown is the underframe detail. The sides, roof and brasswork for handles etc. go into the spares box for use elsewhere one day. Battery boxes and dynamos were stripped from the carflats in the wagon shop conversion process. These white metal parts will become useful weights in other wagons.

Below: Bogies should always be constructed on a flat surface to ensure they are true. Either glass or as here thick board. The cars were also offered up to ensure the MK1 scale length of 64'6" was true here. If a non-passenger carrying Mk1 such as a BG, the length would be 57'6" and four cars would not fit. The chassis and bogies then received a first brush coat of matt black before final assembly.

Cutting and filing the carriage ends. This was softer plastic so a sharp craft knife cutting away from hand and body in a series of cuts was followed by a flex to split then break the groove. The edge was then filed to remove spoil. These useful file sets appear occasionally in Lidl and Aldi or suppliers at shows and are very well priced.

With a 3D-printed hollow section girder in place on either side of the chassis, the load bed is added using square section bamboo cooking skewers cut to size. Matches would serve for OO in the same situation. A safe cutting technique is to roll the skewer with the knife under downward pressure in situ. This will give all four sides and corners a cut as it moves along in constant contact with the knife blade. Then snap the joint and fine trim the ends.

The ninety-two timber balks are then glued in place to form the carflat load bed. The bamboo has differing finishes which take weathering well.

A 3D printer to the rescue. Components of an HO bridge had supports in just the right angle to provide stanchions for the top rail. Some trimming with knife and clippers to size and glued with two-part epoxy resin.

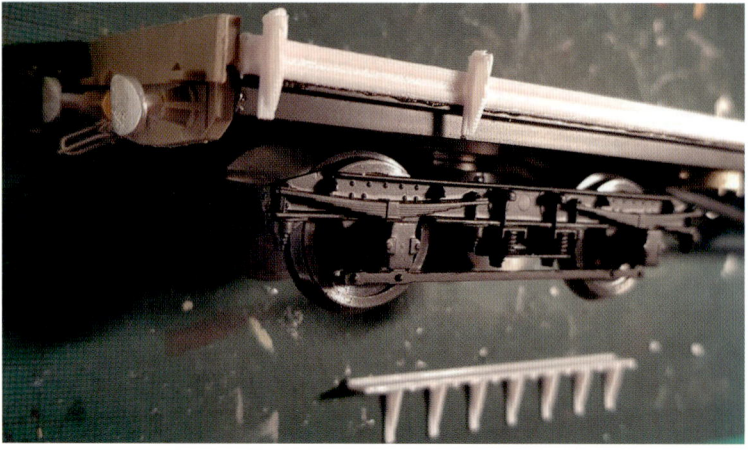

BUILDING YOUR OWN ROLLING STOCK • 75

Some old, damaged Hornby Dublo running rail became the donor for the top rails, adding linear strength to the assembly. These were alcohol cleaned and brushed with a fibreglass pen, then epoxy resin glued into place.

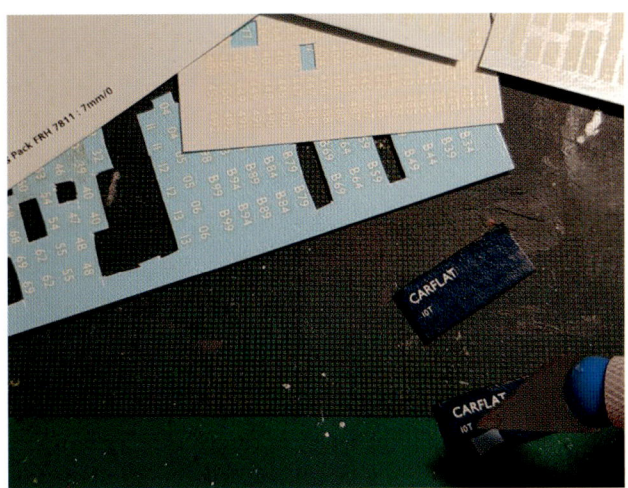

Everyone has their own way of applying waterslide transfers. Here removal from the Fox Transfers main paper is done by a pointed scalpel. The transfer is placed in a saucer of warm water with a drop of washing-up liquid in it. When the film lifts the transfer and backing paper is removed together and the transfer then coaxed into place with the knife tip. If position is slightly wrong, introduce more water to float it and gently dab with kitchen towel to adhere. You can use specialist transfer finishings, but a satin varnish topcoat seems to work adequately.

Something from nothing during viral lockdown. The first carflat B748442 in situ on the author's O gauge Wroxeter Roman Road garden railway. This example is turned out in passenger blue as opposed to fitted bauxite and having looked at images of the prototype has never seemed to receive a visit to the wash plant. These can run as a small factory train or as an unbadged Motorail consist of two carflats and five passenger carriages pulled by the trusty Class 40.

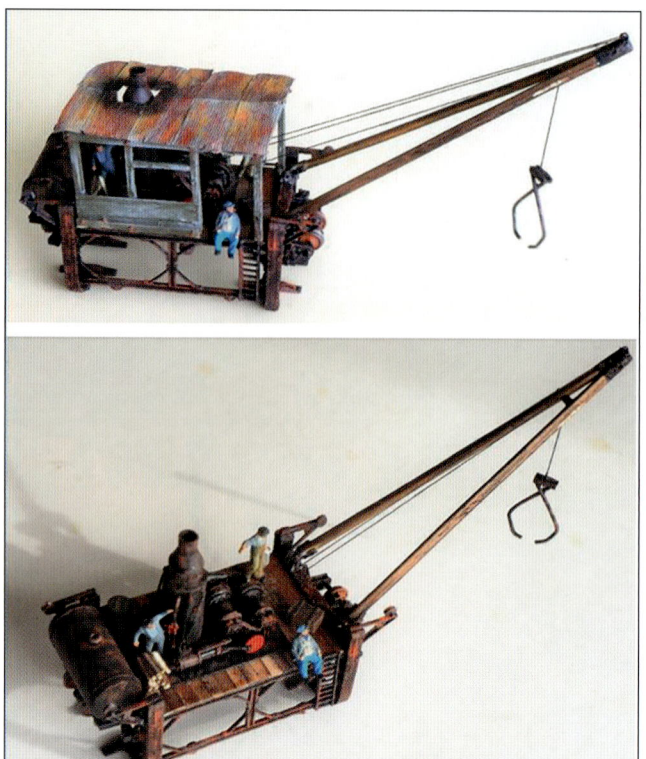

The finished McGiffert Loader, called by some the Cadillac of loaders. The concept was this machine would be able to load up flat rail cars with timber as they passed under. It was self-propelled. With the cab removed the level of detail can be seen. Some parts were missing. The water tank is a pen barrel with plastic card ends. The jib is from bamboo kebab skewers with ironwork from cake wrapper foil and copper tube. Superstructure is from matchsticks and deck planking from another kit stained with shoe dye. Ladders were purloined from the spares box.

Mini project: American Logging Machines

Club member Alan Hancock has a good eye for the American esoteric model kit. Aside from early railcars he is a maker of model sawmills, loaders and ancillary timber handling equipment. Occasionally at the Club we get a donation that includes miscellaneous white metal parts mixed up in a box. Alan had just finished reading up on a loader example so recognised 60% of a kit was present. Enough to give it a go making his own missing parts, all part of the fun.

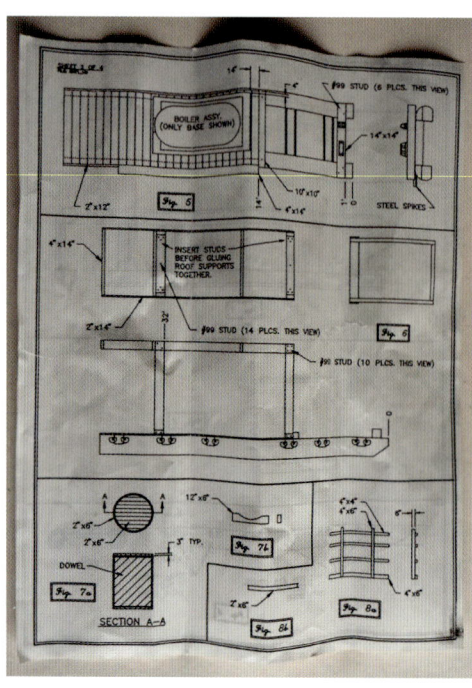

Above, left and right: A 'donkey yarder' for pulling logs. The outsize whistle was to alert out-of-sight lumberjacks that the drag was about to take place. The plans are to a mid-degree of detail and often need interpretation by looking at images of the real thing. You get a box containing many pieces of wood, foil, wires, castings, instructions and drawings. In the UK the barrel would be a solid form with scribing. In an American kit it is made up of rings and strakes for you to perform your own mini coopering activity.

Scratch Building Teak Coaches

We finish this chapter with a visit to the inspirational end of the skills spectrum. Club member Brian Bartholemew is an avid 7mm LNER scratch builder from brass, balsa, wood and plastic card. Some models are built to commission, others for running on the home layout. For example, the Waskerley County Durham snow plough, Gateshead crew van and brake below. Each carriage built is based on prototype drawings and normally will have a bespoke brass underframe and hand-cut timber upperworks with a detailed interior. While we will not be showing how to perform the hot work of soldering brass here, the following pictures have been included to demonstrate what can be achieved once skills have been honed.

Above left: Decisions need to be made as to what level a scratch build will go. Do you cast, machine and fold all your components or do you buy best of breed specialist items where available? With drawbar accessories it is often best to buy in the buffer, coupling and hose assemblies.

Above right: Once you have mastered soldering most of the construction is iteration of what you have learned. Here the battery boxes, generators, stringers V hangers, etc. are part of an underframe kit adapted for the build. However, making your own is an option for most of the components.

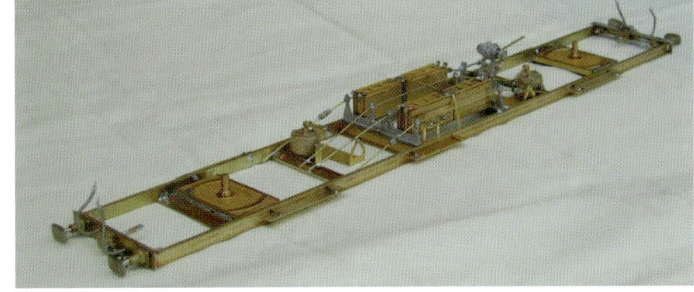

Completed assembly. If you compare with the donor kit used for the carflat the level of detail, skills and finish here are at another level.

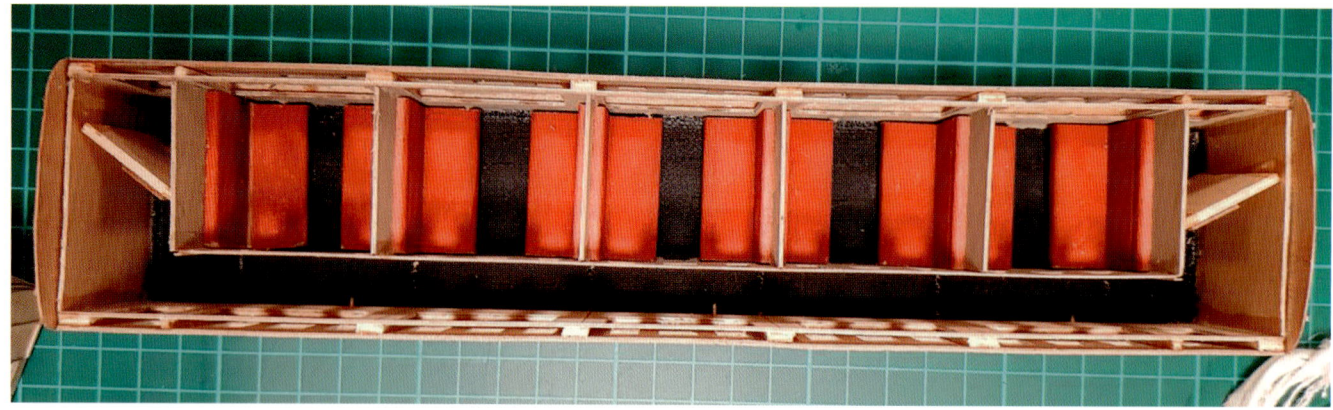

Above and left: Once the sides and ends have been constructed plus brass fitments placed, they are offered up onto an inner frame. This adds strength to the build as the near veneer thickness outers are vulnerable. The sandwich made allows Perspex segments to be dropped into place giving flush glazing. Finally, a series of varnish layers are added, each flatted before the next to give a deep finish.

Below: Pride and joy of the scratch-build fleet is 'The Old Gentleman's Coach' from the film *The Railway Children*. The original Great Northern Railway directors' saloon was built in 1897 and is resident on the Bluebell Railway.

11
Entry-level 3D Printing

BR (SR) car 56 being raised for overhaul from the Waterloo & City line at London's Waterloo Station on 5 November 1950. Below the elevator shaft is a whole network of equipment, unseen from above. The whole hoist assembly here looks like the 3D printing rig for a filament-based printer, creating the carriage with servos and drives below. (*Online Transport Archive – Meredith-155-9*)

What is 3D Printing?

At secondary school the author was set a task to take a contour map and cut corrugated cardboard to represent the area taken up within each contour line. These were then glued together, and the result was a scale simulacrum of the hills of the area chosen. Unfortunately, the Lincolnshire Fens were not chosen for this task – memory serves that it was the quite tricky Leith Hill, Surrey.

Printing in 3D is very much the same concept. An object or component is designed within a CAD (computer-aided design) package and is checked for completeness to ensure all sides are manifold, forming a 'watertight' closed topology. The software also defines where external supports or lightweight infill patterns may be required. You need this otherwise your next step fails.

This design is then put through software that allows manipulation of rotation with resizing of the X, Y and Z axes. It then successively slices it horizontally in memory, rather like the contour map task, but each slice represents the thickness of the hot filament extrusion. The 3D printer takes each of these slices and presents the print head with an X, Y and Z axis coordinate with instructions on whether to extrude molten plastic at that point and how much. The result is a mechanical dance where the hot print head moves over the object, extruding as it goes, as if making a photocopy in filament for each Z-axis layer.

As a domestic modelling technology, 3D printing is still developing and quality improving. At entry level, affordable machines can be purchased easily either as a knocked down kit of components or at a premium, pre-assembled and ready to run. There are many choices to make, and it can become a money pit. You can avoid this with use of compatible freeware, shared 3D model definitions off the shelf from the internet and a single white coloured spool of PLA (polylactic acid) filament. You can start quickly and learn detailed manipulation on the job. Feeling environmentally conscious? The PLA filament is a thermoplastic monomer from such sources as corn starch or sugar cane.

The important thing is to do your homework. Ask someone who has a 3D printer as to their experience and what linking software they use. Look online at Amazon, eBay and other sources to get the reviews and also whether you are about to purchase on the 'bleeding edge' of an untested product. Additionally, is it reduced in price to shift units before an upgrade? Is it something people like, trust and will last technologically a fair amount of time? Saving money up front is not always good if you are only being forced to replace it in twelve months' time.

This chapter covers the basics to get you going on the current technology available and using predesigned print definitions. 3D printing requires the user to know enough about a PC to set up printers and drivers as well as manipulate graphical software for printing. If not your forte, you can consider www.shapeways.com or www.quickparts.com as examples of companies that can produce items for you. Look at online forums such as the Facebook '3D Printing for Model Railroading' or for YouTube tutorials.

All images in this chapter are by the Author.

The Euston Station main courtyard canopy behind the Doric arch in 1875. A delicate feature made from a succession of component prints sourced free online from thingiverse.com and repurposed/resized for use. The real station later replaced this with a much more robust ironwork at the turn of the twentieth century.

The Filament Printer Experience

Admittedly this title sounds like a 1970's rock band name! If you are to part with hard earned money for a 3D printer you do need to know what to expect. This is from our experience of printers owned – there are definitely no sponsors here!

- **It's big**. In order to be able to print a 30cm cube area, the X, Y, Z coordinates are that size plus the framework holding the components in place.
- **It's heavy**. You pay for a fair bit of weight. The printer performs servo-driven hard stops and changes of direction, all with a heavy print head. To keep the rig in place on the tabletop and stop it shaking apart, it needs to be an industrial weight.
- **It's noisy**. Unless you have paid a lot of money for a machine with a sound-proofed cabinet it will be intrusive in the domestic setting. The examples shown here are used indoors, but in rooms and at times that keep them away from the family.
- **There is some smell and dust**. Not too much, but a light ironing board scald aroma permeates. Dust builds over time as a by-product in the area of the printer. Partly this is due to the cooling fans ducting air, thus pulling pre-existing dust and concentrating it.
- **You get what you pay for**. In general, this is true. You can get a cheaper knocked down kit rather than a plug and go version of many of the printers. If you like kits of parts, then no obstacle. Anyone else, the extra money is worth paying.
- **It looks unsophisticated**. The printer itself is practical. Designs are built around best of breed available components rather than design manufactured ones. It does mean many generic replacement parts will fit if things wear out over time.
- **It's frustrating**. To a degree at the start, yes! There is no end-to-end solution. It all feels as if a happy band of industry gurus developed disparate hard and software solutions, some chargeable, some free. You cobble them together to your own best fit. Look online for the latest YouTube guides and recommendations.
- **It's great**. Got this far? Good. Surmount the obstacles and the feeling of deity-like creativity rewards well. Some design not available on the shelves? Chances are you'll find a definition to use. Need four of them tonight? No problem. Pro rata, balancing the product cost saved against the overall price of machine and filament consumed, you'll probably break even. But your resulting model is unique and that shows.

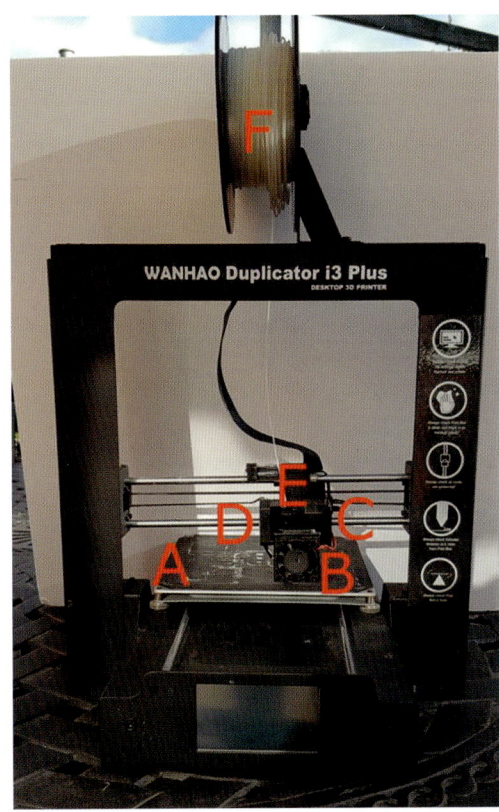

The entry- to mid-level market is ruled by a company called Prusa. They exist in the higher cost bracket but have a quality of manufacture reputation. In having similar features, the Wanhao I3 Plus was chosen as the main printer for Euston's 400 chimney pots, 250 sash windows, roofing ridge tiles plus some doll's house projects before being passed on through the family. It has what is classed as a direct extruder.

A. Print bed. Heated and with a magnetic peelable mat for the physical printing.
B. Hot end. The delivery point of heated extrusion in a crowded assembly. A bit fiddly to take apart to remedy blockages.
C. Stepper motor. Feeds the filament to hot end to create a speed of flow.
D. Heat sink. Preventing the whole assembly heating up and jamming.
E. Filament direct into the top of the stepper motor with open filament.
F. Filament spool above for gravity-assisted feed into extrusion assembly.

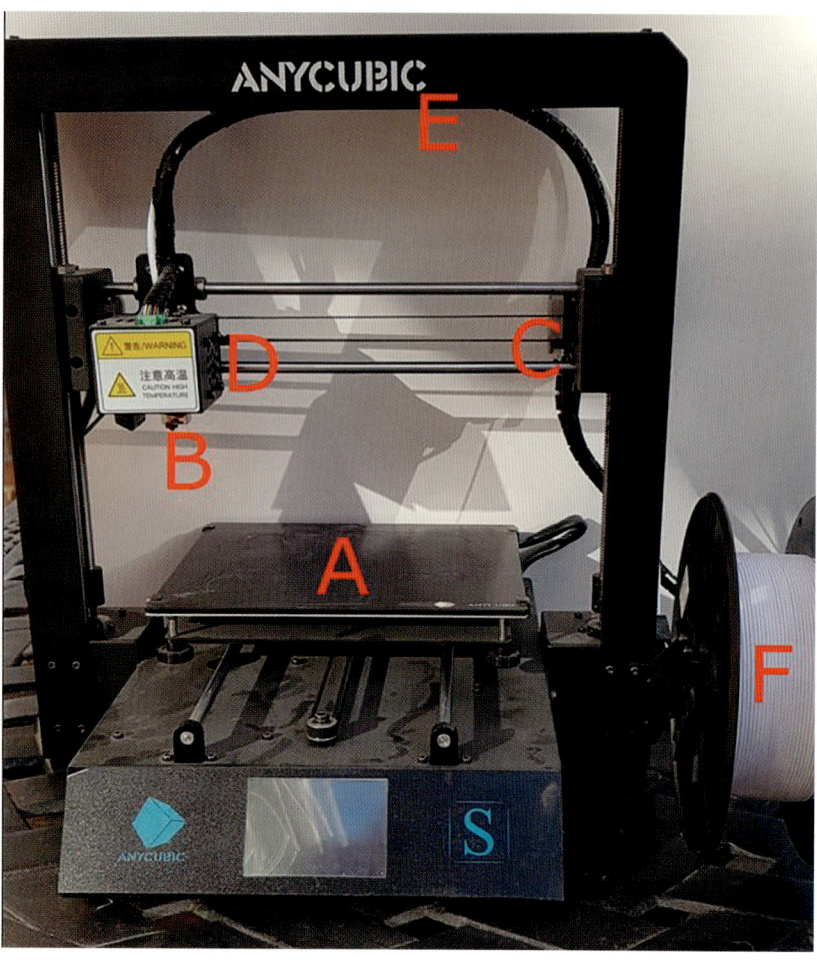

The Anycubic S printer. An entry level printer akin to a rear-wheel-drive car where components are spaced apart and dedicated to function. It is a replacement for the previous printer and has a Bowden-type extruder (step motor is away from the hot end and feeds through a non-slip tube). Downside, it is harder to change filament if swapping of colours is desired.

A. Print bed. Heated and with a painted glass screen which is tacky to printed filament.
B. Hot end. A less crowded hot end where just the heating for extrusion is concentrated upon.
C. The stepper motor drive is chassis mounted, feeding filament through a flexed PTFE tube.
D. Heat sink. Internal, quieter fan assembly.
E. PTFE feed tube for filament supply.
F. Filament spool is side mounted.

Printer Health and Safety

FDM or fused deposition modelling, to give the filament printer the accurate technical name, involves a hot end extruder operating at a minimum 200°. While tucked out of the way and with warnings these printers and children do not mix. Even the heat sink will have an appreciable heat. If the hot end is hot (or cooling down) use remote metal or wood tools if you need to clean overspill of filament away from the nozzle. Work in an area that has a zoned smoke detector covering the unlikely event of a plastics burnout. Treat with care, do not run with broken or exposed electrical/hot parts.

Do not leave the printer unattended. While a long print makes it impractical to sit physically watching all the time, if you have to leave the house then consider pausing the print in the print driver software and let the printer cool in the interim. Resume on your return as it will reheat to operating temperature before restarting This is not ideal as contiguous printing at heat is preferable for strength but, as with a domestic tumble-drier, always play safe.

Above left and above right: FDM printed RAF fuel depot pipework for Andover Road in OO. Small sections are printed, painted and joined to build up complexity. This method of printing is very good for low to mid detail assemblies in flat kit form, but not for high detail and complex single objects. For that, resin printing comes to the fore.

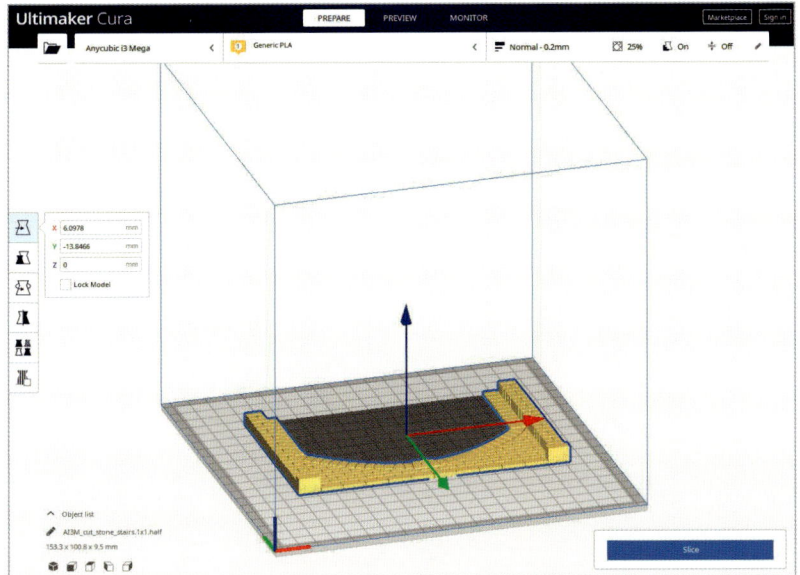

These examples show the progression of a print manipulation and slice using a freeware product called Ultimaker Cura. Here a double tunnel mouth design has been sourced from Thingiverse.com for personal use. Because the eccentricities of this particular printer are known, the first action is to rotate the object 180° on the print bed. This is so that the mass is away from the room wall (it gets a little hotter there and can deform).

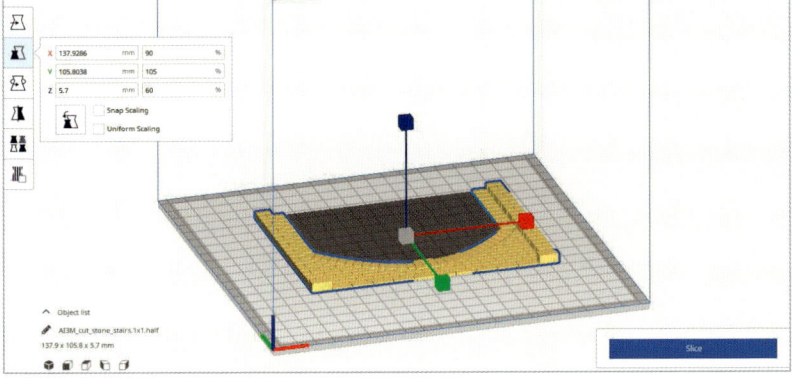

The physical location on the Andover Road baseboard was measured and a 138mm overall width determined. Other dimensions were scaled separately – they can be locked and scaled to the same percentage if desired.

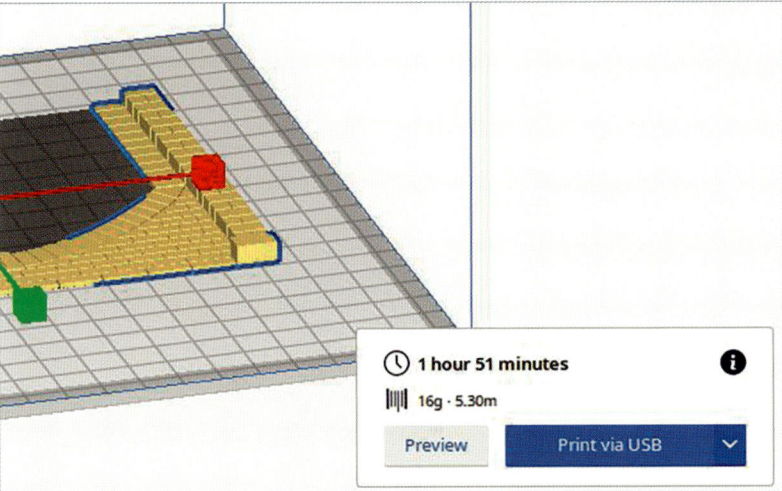

The next action is to get the software to slice the object and give an estimate of production time, filament length and weight of finished object. At 2023 prices a white 750g spool of PLA filament costs £30 and the model is 16g in weight. The tunnel mouth is costing roughly 70p in consumables. Factor in the printer depreciation and, overall, the project cost here is about £1.40

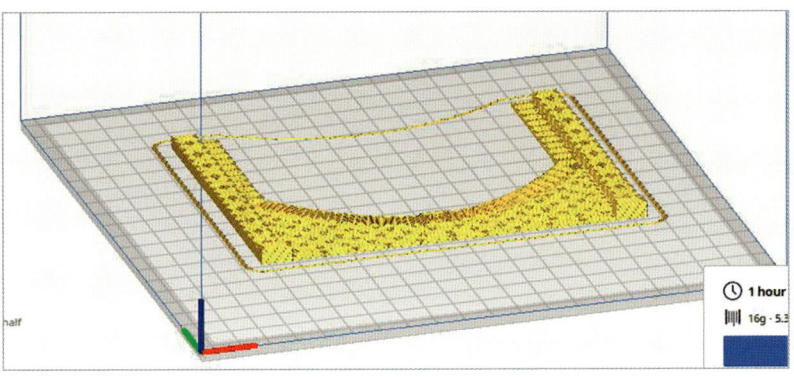

Such designs usually have associated surface textures as a part of the overall design. Here in preview mode blockwork is seen. The surrounding boundary is something that this specific printer does through its driver software. Others may create a thin plastic mat on the printing surface to aid object adhesion.

Commencement of print with the adherence layer going onto the glass hot bed. As you can see these hot beds wear out over time and they can be easily replaced. The aim is to maintain a warmth of about 60° for PLA to ensure the hot filament adheres and there is some plasticity retained during the print cycle.

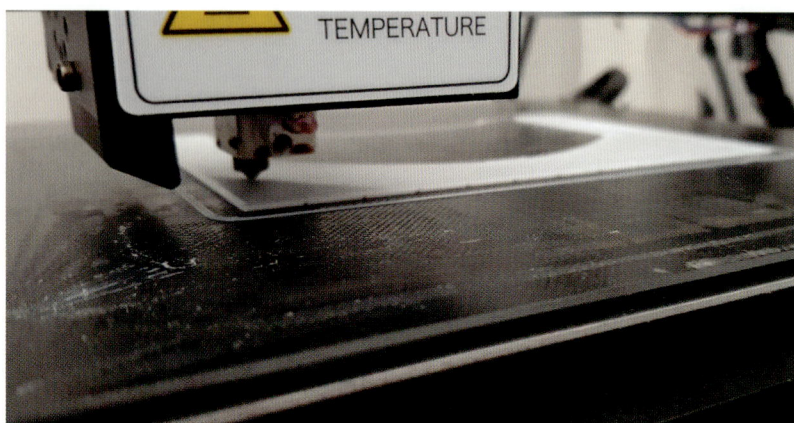

Twenty per cent into the print, probably about layer ten going into place. The extrusion of filament is constant from the 200° hot head and the X/Y axis speed varies according to the density required. It does need the PC always attached and active to constantly feed the coordinate data for layered slice prints to the device.

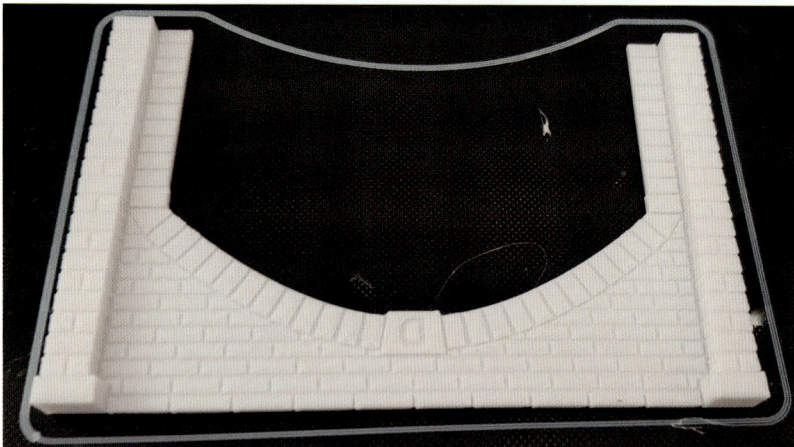

Completed. The printer is now cooling down and the object can be gently coaxed off the glass plate using a quality palette scraper dedicated to the task.

Left: The finished tunnel mouth in situ on the Southampton board of Andover Road which leads to a cartridge fiddle yard behind the backscene. Hornby Class 71 electric unit E3005 peeps through. Right: You are not confined to the horizontal – sometimes it can make sense to move more through the vertical Z axis if your model can adhere to the print bed adequately. It evens out servo and drive belt wear. Here the tunnel embankment retaining wall is being printed.

Putting together a kit of 3D printed parts for use on the O scale Butterwick layout. Again from Thingiverse.com, this time role-playing game scenic items rescaled and repurposed. These originated as a white PLA filament and were then primed and surface painted ready for offering up into position. Stopping a print short using the abort option left intentional stumps on walls to represent wartime cutting of iron railings for salvage. Only one panel was taken to completion.

A display of Victorian chimney pots on the Isle of Wight which was taken as source material for projects. Club members often bring prototype component images to active build projects to use as guidance.

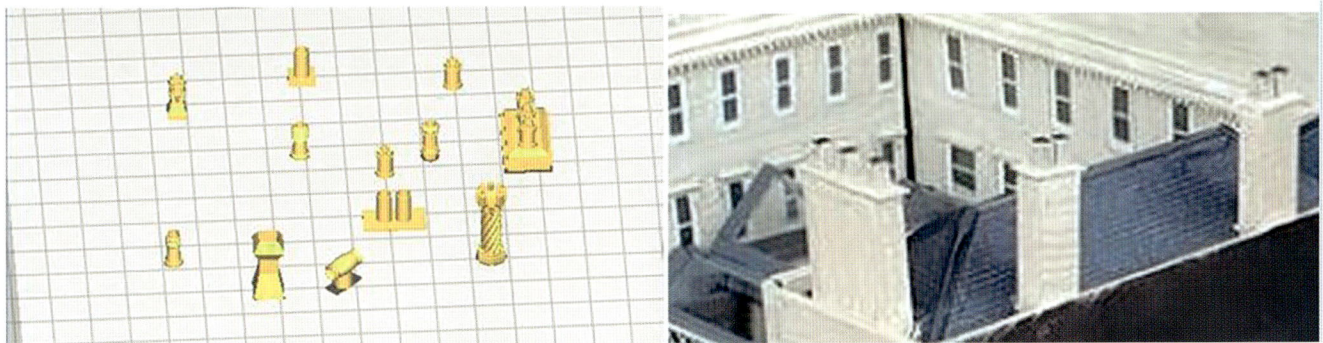

Chimney pots sourced as free STL (stereolithography) models on the web and used on the 1875 Euston layout as a time saver. For example, the squared pot on the lower left was truncated to become a Euston station trapezium specialist pot and the candy twirl one enlarged became the station pots on Butterwick.

Solving Problems

Printers normally come with a small set of tools and a guide to solving simple problems. A quick web search can often gather solutions from those online who have suffered first and altruistically uploaded for you.

Common issues are the filament slipping in the drive gear threads or getting jammed in the carrier pipe (normally a soft PTFE type non-slip pipe to nozzle). The filament can also snap if the feed drum gets bound up, so you end up trying to manually feed, pushing the breakage through until the pull of the gear teeth is felt and feeding starts again.

After extensive use, the heated nozzle and element can begin to burn out requiring replacement. Not a major thing, but harder than replacing the toner in a modern printer.

Printer plates can become damaged or lose their stickiness over time, especially if you have not been moving print start points around to even out wear.

Considering that the technology itself is comparatively new, and the emerging 3D industry is making use of many generic parts put together in different configurations, it all hangs together quite well.

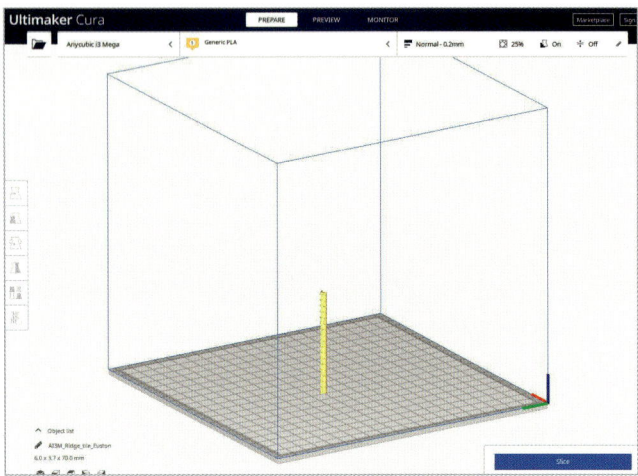

A ridge tile design was sourced on Thingiverse.com and saved to PC as an example to be manipulated. In its native format from original it was thought suitable for immediate printing. To avoid the requirements for supports in the print activity it is oriented long dimension upright. While the glass plate print bed is tolerant of a small adherence footprint, this was regarded as marginal.

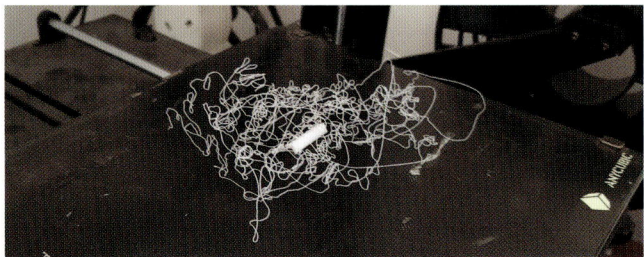

Everyone runs into this messy tangle occasionally, when the adherence on the print plate fails or there is an incorrect design. If unattended the printer continues to extrude hot plastic to where the actual printing should be, but is now in mid-air. The Cura software running the printer has an abort option to terminate early and cut your losses. You have to be there to take advantage of it. Here the design was too narrow for the tall V-shaped print of the ridge tiles and failed after a quarter of the print time had elapsed.

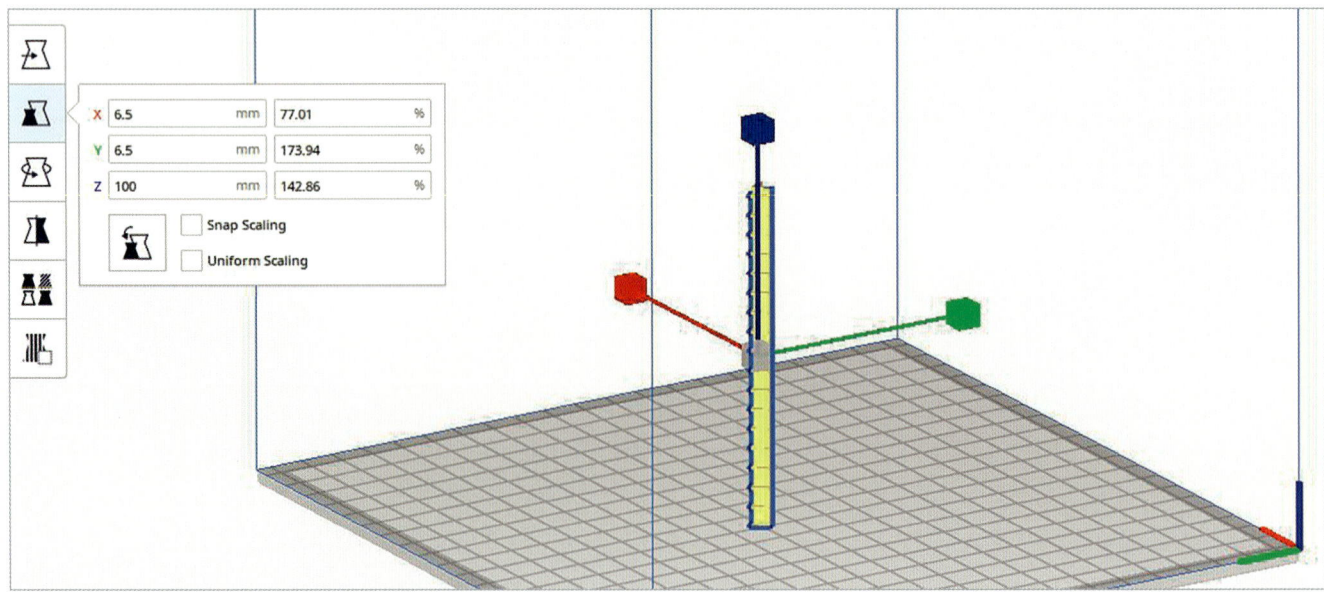

Back to the literal drawing board for a resize of the item. Length was maintained but the X, Y dimension was changed to increase the footprint. The scaling relationship was removed at this juncture to avoid all dimensions changing. The ridge tiles are now deeper but not longer per unit.

Right: Home and dry. Well almost. If cutting to length is required, it is often wise to use small clippers or wire cutters as the PLA plastic is very hard and often brittle. A pair were used here to cut the ridge tiles to length, as well as the printed guttering. Admittedly at this point it also garnered the interest of Lucy the kitten who promptly knocked the chimney pot off this building for Market Obthorpe. Such is to be expected when using the dining table!

Below: Oh, it's a mess. Occasionally the printer filament extrusion will just not adhere to the plate. It can be because of a dusty plate or manual readjustment of the print plate being required as the level setting nuts can shake loose over time. Although frustrating it is easy to remedy.

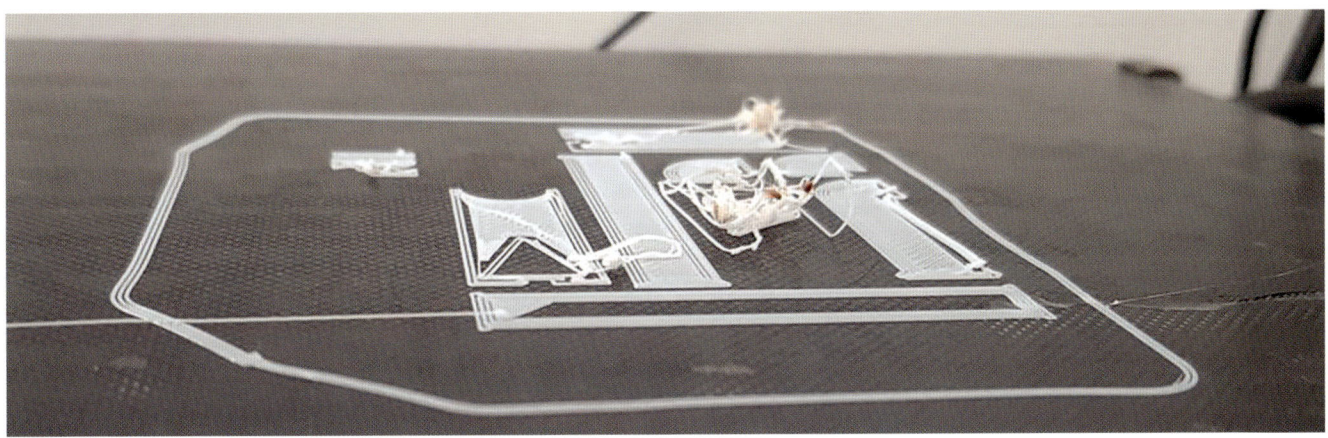

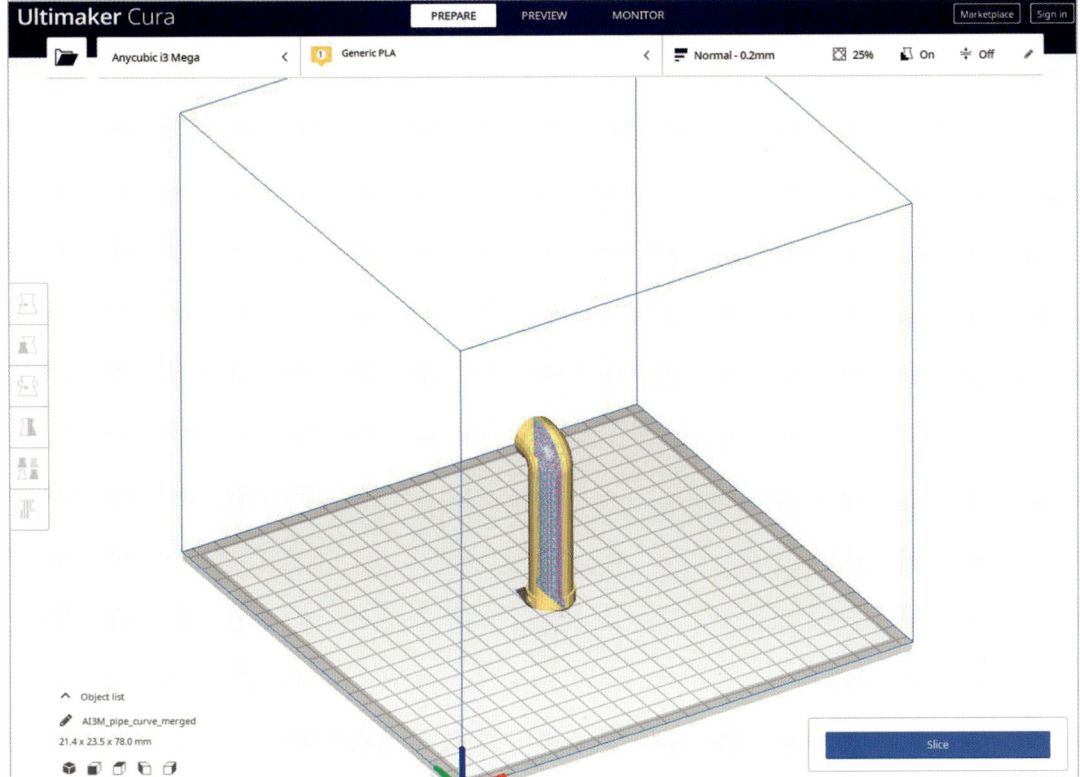

If your 3D model is not manifold, this implies that the model is not 'watertight' in some form. In this example two halves of a fuel depot pipe designed to be printed flat and then manually glued have been merged to form an upright single print. As part of the merge, some flatbed texture has not meshed well, leaving gaps. In this example it was not drastic and was solved easily. In other models you could end up with a texture hole requiring tidy-up at the end. In the worst case you may attempt to print in an area where plastic has already been extruded which could potentially damage your printer.

Mini Project: Making the Loos

While there is much available for the OO modeller off the shelf or easily adaptable, the O gauge aficionado still has to step back, plan and make bespoke items in many areas. Butterwick needed a gent's loo front of house and in the open air. Perhaps a mundane requirement but if you've ever been trapped on a station when they're closed you get to appreciate them. Therefore, this was chosen as a mini 3D printing project based on seeing the prototype masterpiece when attending a model show at Doncaster racecourse.

Having done this it has generated a lot of interest as there are very few traditional open station examples surviving outside of preservation. Certainly not many O gauge examples at the baseboard's front edge.

Euston in OO has the cast-iron urinal from Slater's sitting in the ornamental gardens.

The Twyford Adamant urinal below the 1881 listed grandstand of Doncaster racecourse. An architectural masterpiece. Colleagues outside in the frosty air thought I was a bit crazy nipping into a building unused that day.

The author's own attempt, inspired by Tyford Adamant, albeit reduced in size for the branch line terminus. You can see the striations that filament printing leaves behind. The smoothed upper joining pieces were abandoned as being too difficult to create in 3D mode. Creation with DAS clay was attempted but the result was a cluttered look from above so deemed not needed since it looks right without. Unless you are following a tight prototype do not be afraid to adapt, often only the model builder knows the exact original target.

Above left and above right: Station loos have surrounding vanity walls. Self-printed brick papers, some tile bits from the Metcalfe scrap box and *voilà*. My wife at this point suggested leaving it like this and crediting the San Francisco Museum of Modern Art (MOMA) where she saw the original artwork urinal of Marcel Duchamp.

12
3D Resin Printing

With the SS *St Helier* in the background, here is the transhipment quay at Weymouth under construction on 23 June 1951. With the contractor's crane in use it feels like a 3D printing rig. Constructional projects and detailing can be greatly aided by the 3D printing technologies available. Create your unique requirement, use, and manipulate an existing design or purchase a detailed example in digital format. (*Online Transport Archive – Meredith-200-8*)

Resin Printing Compared to FDM Printing

Club member Brian Norris has been spearheading research into the realm of resin-based 3D printing. This chapter is based on his experiences of the method, and all images are his own.

FDM printing

FDM is great and works by taking plastic filament, heating it up and squirting the molten plastic through a small nozzle onto a print bed akin to icing a cake, printing from the bottom up to the top, a layer at a time.

The bed and the head move to ensure that the layer of filament is laid in the correct place for each layer before moving the print head up a layer and repeating the entire process again.

All the above requires very clever software to control everything from temperature sensors to stepper motors that drive the print head and the printer bed very accurately, and of course the software to read the G-code of your design.

Your slicer program, often distributed free by your printer manufacturer, will produce the G-code for you.

Good as modern printers are, the finished article will still show striations or fine lines. Depending on what you are making this may require further treatment (e.g. a light sanding or use of paint containing fillers).

If you choose to print more than one object at a time the print head has to travel and deposit filament. Therefore, every extra object increases the time taken to print.

The commonly used 0.4mm nozzle will, as a rule, produce a layer thickness of up to 0.32mm or 80% of the nozzle diameter.

Resin printing

Resin printing, on the other hand requires far fewer moving parts as the resin used is UV sensitive and is cured (hardened) one layer at a time by exposing each layer to UV light generated by a screen. The screen is like that used by a mobile phone and sits under the resin tank. The resin is held in a tank with a transparent bottom (FEP film) which sits on top of the LCD screen. The FEP film, commonly just referred to as 'the FEP' is a sheet of special transparent plastic which is flexible and resistant to UV light.

The only moving part is the threaded rod at the rear which, by rotating, lifts and lowers the build plate assembly. The print sticks to the build plate which gradually moves the whole model upwards, between 0.01mm and 0.05mm, one layer at a time.

This results in a much finer finish on the print than that available using FDM printing although the print lines can still be visible.

A finishing curing step needs to be performed to produce a hardened finished object. The UV light of the printing process produces a layer by layer linking. To get an item that can be robustly handled and painted in a short period after creation, a UV cabinet is ideally required.

Once you have tested a print you can multiply it on the build plate if there is space. The print time will not be a multiplier of the number of objects printed as each layer is a UV exposure with pixels either on or off.

The resin printer referred to in the this chapter is an Elegoo Mars 2 Pro which at the time of purchase was top of the range. A good choice for the home user, and entirely suitable for the modeller in N through to O scales.

Newer more advanced models have subsequently become available from a range of suppliers.

Resin Printer Health and Safety

The resin and other chemicals you may use are potentially harmful to both you and the environment. However, follow the instructions on the labels with care and all will be well. As examples, here are photos of the warnings on the resin and isopropyl alcohol (IPA) bottles.

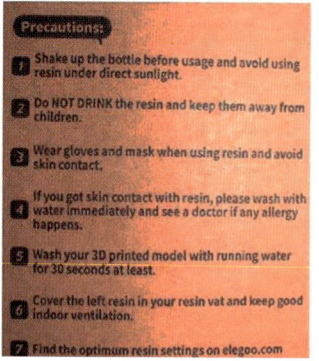 

Above left: A resin bottle example. Contact between skin and resin can cause skin irritation and you should always wear gloves. Resin gives off fumes and while some printers come with activated carbon filters, good ventilation is strongly recommended.

Above right: IPA Bottle example. IPA gives off strong fumes and exposure can cause nausea and drowsiness. Good ventilation is essential. Always print with the machine cover closed. That cover is not there just to keep external UV light out of your resin tank, it is also to protect your eyes from the UV produced by the printer.

The Printer and Tools

The Elegoo Mars 2 Pro is used here to show process since it is the printer owned and operated by Brian. Since buying the printer the industry has matured quickly and many other printers have been produced. If you are considering buying a printer, it is worth spending some time researching them on the internet.

The two main factors you should be interested in are the size of the print bed and the definition of the machine in terms of layer height and pixel density, as these will determine the fineness of the finished print. Different resins are available and the general price of them may also determine decision making.

Guide to the Process

The print prints from the bottom upwards but unlike an FDM printer, it is printed inverted. This means that the bottom of the print is stuck to the build plate which slowly rises, gradually printing higher and higher layers of the model until finally the top layer prints and the whole print is withdrawn from the resin bath.

Most printers will come either with free software on a memory stick or available as a download from the internet. Either way, be sure to make a secure backup of all the files. The Mars Pro 2 came with a memory stick containing all the files needed for the operation of the printer together with a mains lead, transformer and a variety of tools.

1. The main printer body and touch screen with USB stick inserted at bottom right. The body includes an extraction fan and filter.
2. The build plate, normally horizontal but in this picture suspended in draining position.
3. The resin tank with FEP (non-stick film) sheet already installed. Spares were supplied.
4. Rubber-edged, red cover to exclude ultraviolet light and contain fumes. This sits over the main body of the printer while printing is in progress.
5. The Z axis screw for raising the build plate.
6. A range of tools including, L to R: plastic squeegee for gentle cleaning of FEP sheet; side cutter for removing supports; soft brush, useful for rinsing out the resin tank; metal scraper for removal of prints from the build plate.
7. Selection of resins. Two shown, one water soluble, the other requiring IPA. Both types come with health warnings which should be well heeded.
8. IPA used for washing out non-water-soluble resins. This gives off a lot of fumes and should be used in a well-ventilated area.

Not shown – protective gloves, fine filters for filtering unused resin back into the bottle, mains lead and transformer.

More modern printers have dedicated monochrome print screens which are said to be more efficient than the older LCD screens used in smartphones. The older screens emit only a small percentage of the UV light needed to cure the resin. Monochrome screens should print faster, and as screens improve so will definition.

 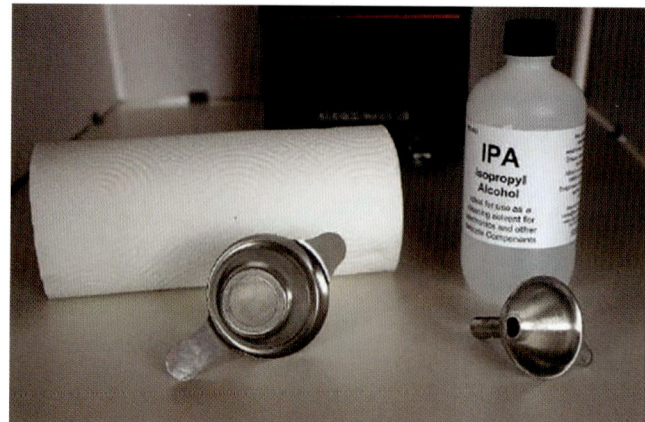

Another view of the printer showing the printer plate and resin tank demounted with the white USB stick and draining attachment in the centre.

Other items you will need: fairly quickly you will find you want more gloves, filters, IPA and varieties of resin. In addition, you will want:
- 'tanks' (e.g. margarine/ice cream tubs) for washing prints and utensils
- funnels for holding the filter papers
- rolls of paper towel
- fine sieves, both metal and paint sieves, seen later.

Beginning the Print

Having set up your machine per manufacturer's instruction, it is time to start printing. Copy your print file to your SD card, insert into the printer and turn on. If you are sure you have zeroed your print head, fasten it down, plus the resin tank. Always ensure that you wear gloves.

Fill the tank with sufficient resin. There should be a 'max Level' indicator on the tank. Start the printer. Put on the safety UV cover. Listen and watch for a few minutes and you will hear a happy machine.

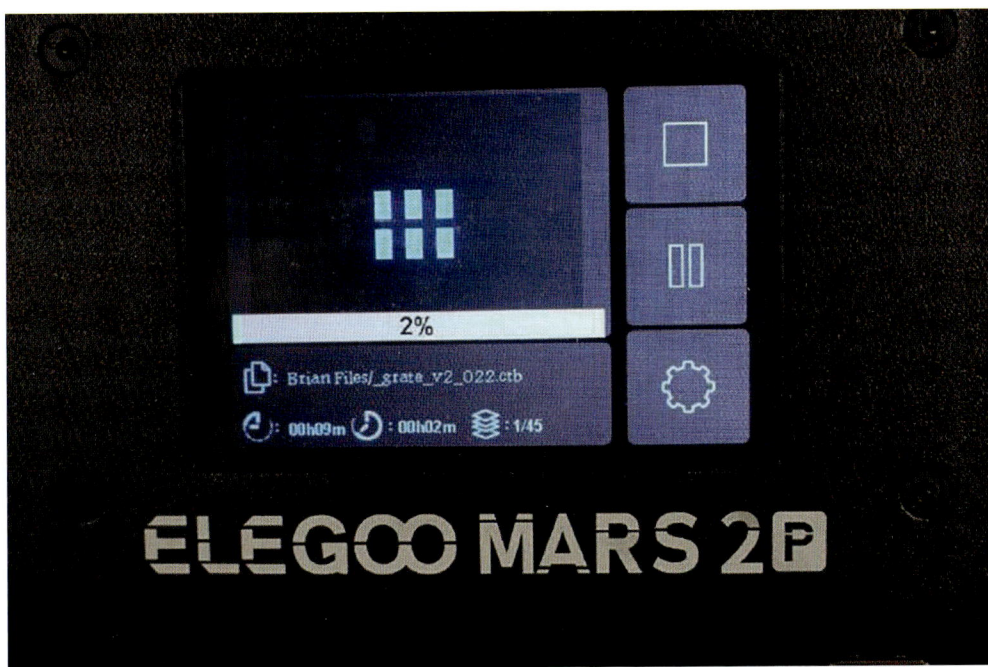

This LCD screen touch sensitive control panel has information on the left and controls on the right.
  Among the information is:
- a picture of the layer currently being printed
- the percentage of the print, finished
- the file path on your SD card
- print time remaining
- print time elapsed
- current layer number out of how many layers.

Clean up. Resin seems to get everywhere. Cleaning the machine is important as it will directly impact on FEP replacement frequency. Leaks and drips must be thoroughly cleaned up swiftly to avoid contamination of surfaces. There should never be resin between the FEP and print screen. Remove the build pate first with the print and put it into a cleaning tray. You must prevent resin dripping on the LCD screen. Unless planning to do another print immediately, return the residual resin to the bottle. Some users put used resin in a different bottle, so that for important prints they can be sure of perfectly clean resin. Sieves can easily become clogged, and it has been found best to use paint strainers available from your local garage or via an internet search. Sieving the resin stops small bits of set resin from getting into the tank. In the image, a paint strainer with a 190-micron mesh has been cut down to better fit the green funnel.

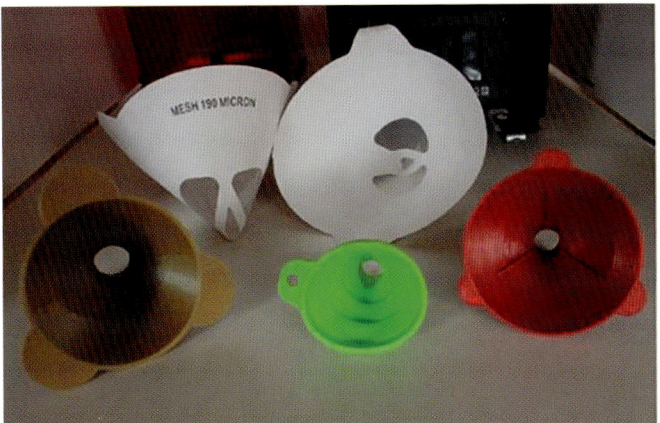

Above left and above right: Pre-prepare cleaning tanks, filters, sieves and utensils to assist with cleaning up at the end. Put gloves on. At the end of print the head rises as far as possible. Place the print head into the draining position. Resin is more viscous than water and hence has a slower drain rate. Put the outer UV cover back on while draining. As much as possible, avoid direct or bright sunlight. Note the inadvertent spill on the left of the resin tank. This had to be cleaned up immediately since the resin could seep between the tank and the LCD screen.

Above left: Squeegee out the resin from the tank FEP sheet using a soft scraper to avoid damage, then wash thoroughly with water or IPA as appropriate, using the brush to dislodge resin around the sides. Check that there are no pieces adhering to the FEP. Here there is the remnant of a failed print resisting being brushed off.

Above right: This piece of resin resisted and the soft scraper was resorted to. There is a single piece of paper towel under the FEP and that, in turn, is on a smooth, hard surface. Do not dent the FEP or it will require replacement.

Important note. The resin may say that it is water soluble. All that means is that water will serve to dissolve the resin allowing you to clean, *but* the uncured resin is highly toxic to wildlife. The water/resin mix should never be poured down the drain. Put the mix to one side in a sunny window and the resin will settle out and cure to a paste for solid disposal. If in doubt, check with your local authority or environment agency.

Curing the Print

The resin is cured by being exposed to UV light from the printer's LCD. When the print is removed from the printer it will still be slightly soft and malleable. This helps in the removal of the print and the removal of the supports, but the print needs to be fully cured before paint is applied. The cheapest and easiest way to cure the print is to put it on a sunny windowsill for a few days. The thickness of the print will determine the length of time required to cure.

Adopt a similar routine for the print with the exception that it also needs removing from the build plate. This is where the metal spatula comes in but beware, it can easily damage your build plate which is a costly replacement part. How well the print sticks to the build plate will depend on the initial exposure time for the first layer. Reducing this a little may make your print easier to remove. Too much may cause it to stick to the FEP rather than the build plate. Experiment carefully. This oil barrel is the example project that is covered in the later design chapter. Supports here may break off during the removal of the print from the print head. Those that do not are stripped off using micro clippers.

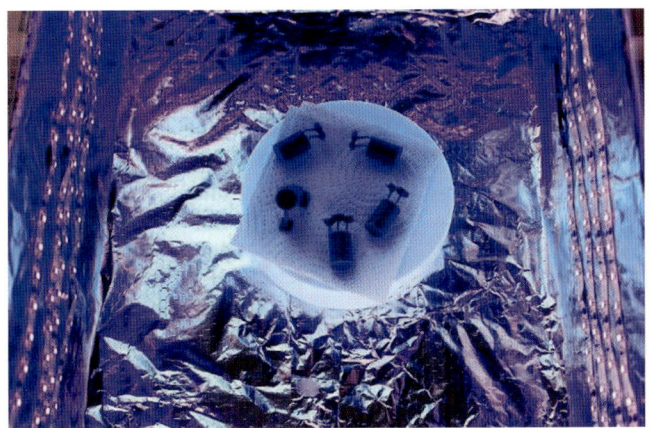

Alternatively, you can buy or make a curing box. Line it with aluminium foil and run a reel of low voltage UV LEDs around the sides. A rotating table can help ensure that all sides get exposed. The critical wavelength for the LEDs is around 405nm and a reputable supplier of LEDs should carry this information.

Printing Failures

Sooner or later you will experience a failed print. Here is a failed print where it has stuck to the FEP rather than the build plate, producing a blob.

There are many possible reasons for failure.

- **Insufficient exposure on first layer**. This was probably the reason for the failure above where the outline of the support bases are clearly visible. This failure was on a test print for a barrel with a first layer exposure of only 25 seconds rather than the 35 seconds recommended by the software.
- **Dirty or worn FEP** (the non-stick film). Cleanliness is key and any odd resin flakes stick on the FEP (they can be highly charged with static electricity and difficult to clean off). Before printing, invert the tank and clean both FEP and tank with a paper towel dampened with IPA. Then hold up to the light to see if you can spot any defects.
- **Dirty resin**. You may have strained your resin, but it is tempting to do prints 'back to back' to save cleaning time. However, if any part of the previous print broke off or was an 'island', it could be floating around in the tank.
- **Dents in your FEP**. It is very easy over time to accumulate scratches and dents in the FEP.
- **Inadequate supports**. If you are printing something thin, it can flex and while it may print completely it can come out curved.
- **Overhangs**. Imagine you are printing a cube and for whatever reason have decided to tip it over at an angle. You now have overhangs – shapes that extend outwards without any support other than that of the layer below it. The maximum recommended overhang is 45° as that allows 50% contact with the layer below. As far as possible try to avoid an angle of more than 30° and provide supports where greater overhangs are unavoidable. As in all things, there is an exception to this – see 'cupping', below.
- **Trapped resin**. This is not a printing fault but a design stage flaw. Having an area inside your print which contains liquid resin is not a good idea. If the print cracks, resin will leak out with undesirable results. Generally do not hollow small-scale prints. If you must hollow your print, ensure you provide two drain holes. One for the air to get in while the resin drains out of the other. In this way, you will avoid cupping, described below.
- **Cupping**. Imagine you are printing a cup. Print with the base to build plate and you will soon have a large area of liquid resin contained within the cup and the rim will need to be pulled off the FEP for each new layer. The rim of the cup will remain immersed to start with and as the numbers of layers increase, so will the weight of the resin being pulled. Eventually one of two things will happen. Either the rim of the

Failed print, showing the resulting blob.

cup exits the resin or the layer adhesion fails. Either way a mini tsunami occurs with potentially disastrous results for the print and the printer. Try it with a glass full of water being pulled off the bottom of a bowl. Turning the cup through 180° might improve things but you again face the issue of what happens when the print is finished. In this case you have a cup stuck to your build plate, which is full of resin. Getting it off will be messy. Turning it through only 90° is probably the best answer.

- **Sometimes when printing multiple identical items, one may fail**. There may be no obvious answer for this but before putting it in the bin, think about having a scrap yard on your layout!

Finishing and Painting

As with all things, before painting examine your model carefully for flaws. Usually this may just be the odd bump or dimple where a support was attached and a little judicious sanding with fine sandpaper or use of a filler will resolve the issue.

Normally when using the spray booth with a can of spray paint there would be more items on the lazy Susan which is turned manually to apply an even coat on all sides. An undercoat of grey primer will usually show up any areas in need of either filing or sanding and a further coat of undercoat should then leave the way clear for application of a finish of your choice.

Holding the piece during painting can be an issue because of the small size of the parts. It depends on the piece as to the approach used which can range from a blob of Blu Tack, tweezers or drilling a small hole somewhere and propping it up on a cocktail stick.

Once happy with the finish, it will have any transfers applied and a coat of satin acrylic to impart a faint sheen. In the next chapter we cover the actual technical design phase of the oil barrel used in the example print process.

13
Computer-aided Design

Sulzer Class 24 D5001 ticks over at the head of a parcels consist at platform 5 of Blackpool North station in 1966. This elegant, canopied part of the station had a temporary roof refurbishment in the early 60s designed to last just four years. In the end it was demolished in 1973 after which all traffic would pass to the excursions station next door. If you deconstruct this scene there are many elements that could be 3D printed as they form a kit of parts or unique items. Railings, columns and spandrels, the bridge and the lamp posts could all be CAD designed and printed using either FDM or resin techniques. Alternatively, if CAD is not your thing you can seek free designs online or purchase detail model definitions. (*BR (M) Derby*)

Computer Design Project: Oil Barrel

Throughout this section the references to computers and programmes are based around a PC running Windows 11 images and experiences are from club member Brian Norris.

The same range of software available for designing for FDM printers is available for resin printers, the only proviso in both cases is that the software must be capable of exporting the file in .STL format (stereo lithography) which is the format that the slicer programmes can understand. Examples includes Blender, Design Spark Mechanical, FreeCAD, AutoCAD Fusion 360, OnShape, OpenSCAD, Sketchup Make, Solidworks, Solid Edge Community Edition and Tinkercad. This is not an exhaustive list but most of these are entirely free and/or have Windows, Mac and Linux versions available. Many programs also have paid professional versions available with more complete tool sets. A search on the internet will quickly take you to the download site for each one.

Reference sources. The first thing to do is find reference material, ideally by going out and taking pictures or alternatively making an internet search. Internet searches are fine, but you will need to know the size of the original item.

Research phase. An old oil drum being used for burning rubbish. It appears to have had two metal bars welded to it at the top and a hole cut into it on the right-hand side. The ruler at the bottom gives a sense of scale. Beer barrels on the other hand, while similar in purpose, have a distinctive design in that their rims are much higher than the lids and they have handles cut through them for easier handling. While the kegs have a similar flattish outline, the beer barrels are more pot-bellied.

Ideally, for a reference picture you need to have a front and a side view, and it needs to be as clear of other parts of the picture as possible. This oil drum has been extracted from the picture above using Photoshop. Note the presence of the ruler in the picture. This is purely to give an idea of size and aid rough calculations. A height of 900mm by width of 600mm is suggested. Note these are not precise measurements, simply the closest easy numbers. Part of the reason for choosing to model an oil barrel was that it is simple to draw as it is just a cylinder with thickening at the ends for the rims and ridges twice along the body. Many oil barrels have additional smaller ridges, to provide greater rigidity and strength.

COMPUTER-AIDED DESIGN • 99

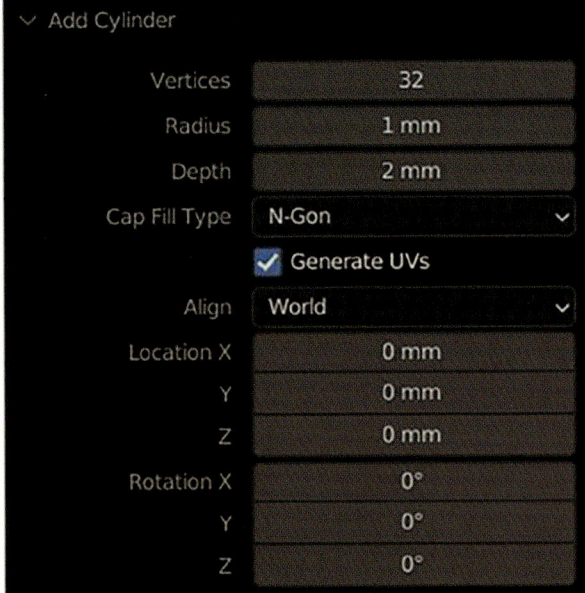

Above left: **Production**. The drawing is produced in Blender (www.blender.org), simply because it is a familiar design product. Stage 1 is the import of the reference pictures. Normally, when importing reference images, look for front and side views. For a cylindrical object, these are identical apart from the lettering on the barrel and only a front view is necessary at this stage. Here is the image being scaled to the cylinder which looks like a rectangle as it is in front orthographic view.

Above right: Here is the image 'behind' the cube. The background picture was taken at a slight angle so there is a degree of perspective parallax involved and this is a 'best fit'. Then Stage 2 – adding the cylinder. Most drawing packages draw circles by using a series of straight lines or segments between vertices. Blender does this here with default set at thirty-two.

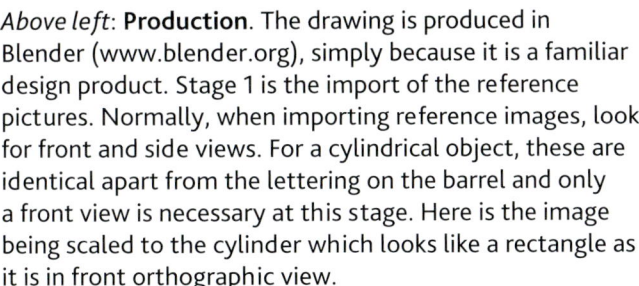

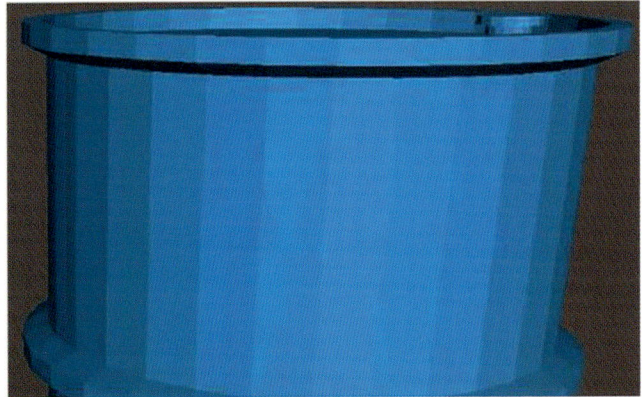

Right: The number of vertices can be altered from three. If too few are used it can produce an effect like that shown here where the segments are clearly visible. If too many are used, there is a risk that Blender will 'fall over'. Above is the object with the default number of vertices, a bit blocky. Below is the cylinder sized to the image. Smoothing by use of sixty-four vertices was chosen and, while the segmentation is still visible it will be less so once scaled.

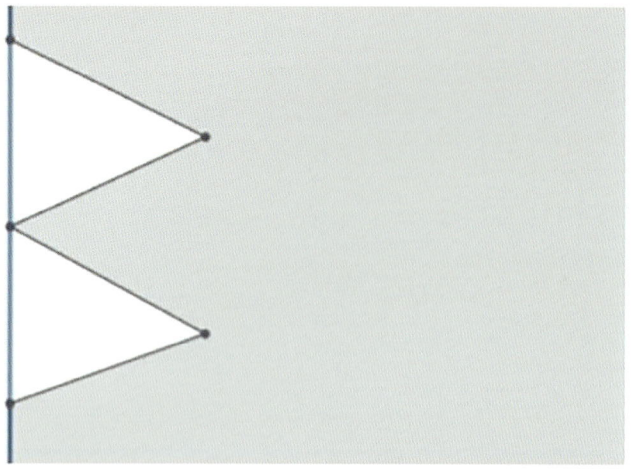

Now to Stage 3, the sides. – adding in the ridges of the barrel. Referring to the reference image there are two ridges and if viewed in cross section these would be in the shape of a 'W' sideways on.

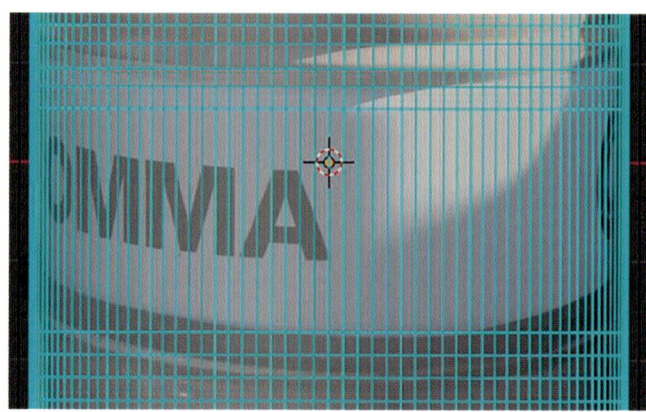

Lining up the rings with the prototype image. Note that because the background picture was taken at a slight angle to the barrel, the rings match here in the centre but not at the edges.

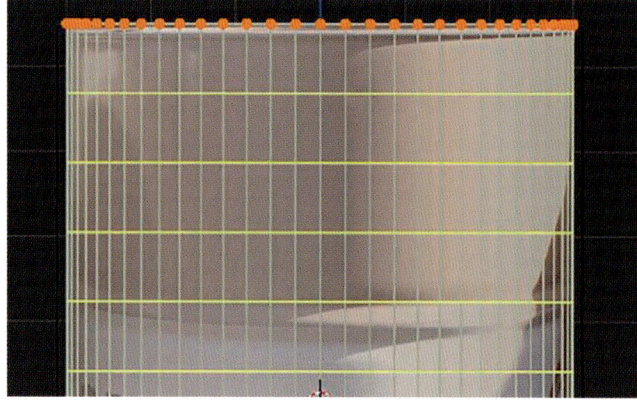

Next for the drum: for each crease at least five rings are required around the barrel. The top and bottom rings act to hold the sides of the barrel in place, the two minor ones will be of a smaller diameter to pull the barrel inwards and the middle ring gives us the peak of the central ridge.

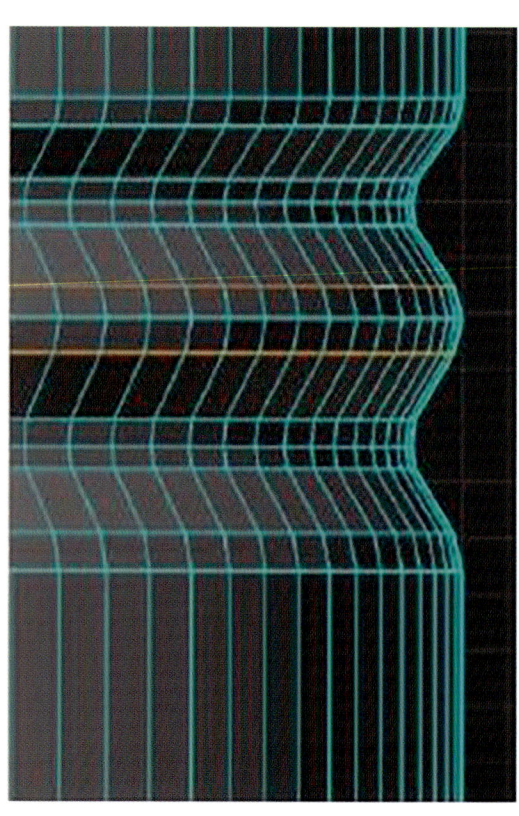

Left: Scaling the rings. Here are the two smaller rings being scaled on the X/Y plane together to make sure they are identical. *Right*: This gives quite sharp edges to the ridges which in reality are quite rounded. Adding bevels to the ridges makes them look more realistic.

COMPUTER-AIDED DESIGN • **101**

A similar treatment is applied to the top and bottom rings.

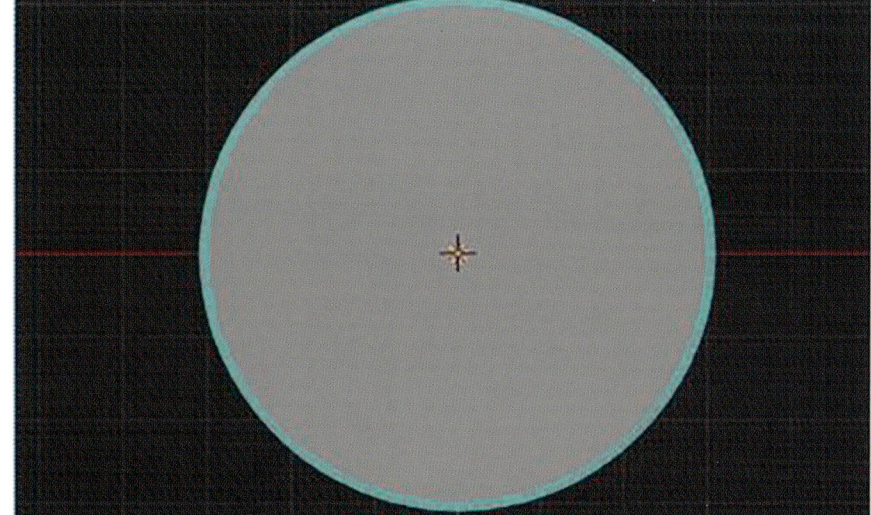

Above right: Nearly there. Stage 4 – top and bottom Because of the way Blender draws cylinders, the top and bottom of the barrel are single faces. The extra blue outlined faces seen here are parts of the rims. *Middle*: On a barrel, the tops and bottoms are inset to make the rims. The top of the barrel is lowered sufficiently so that the barrel openings do not get accidentally knocked.
Right: The top of the barrel with rim completed.

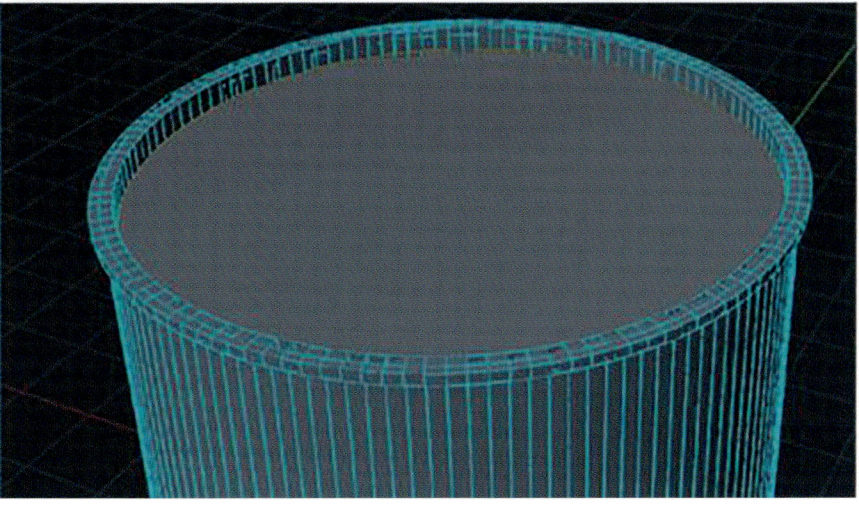

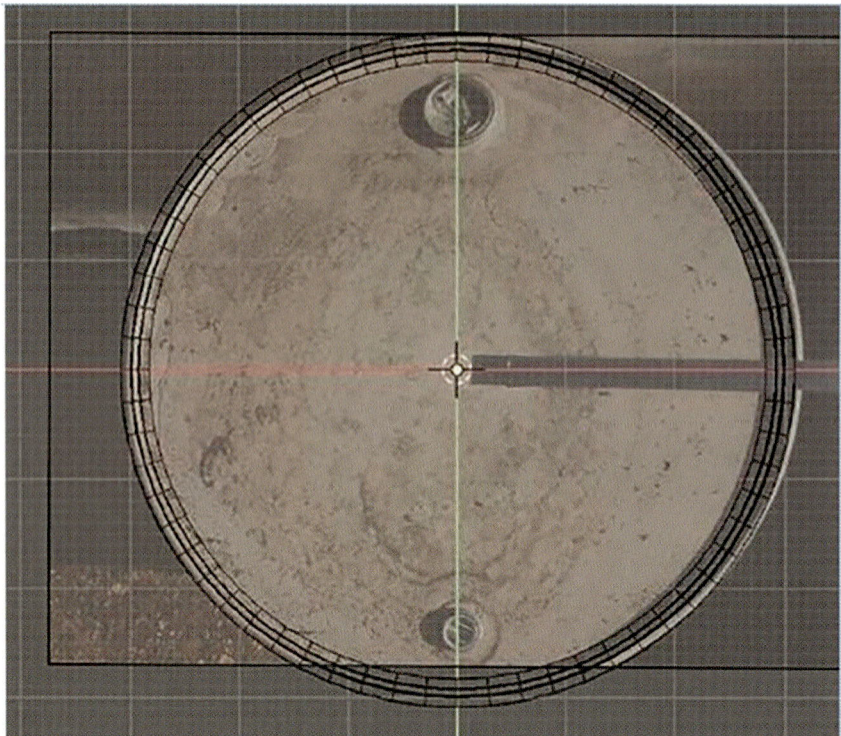

Above left: For reference purposes again, here is a picture of the top of the barrel with a wireframe view of the model barrel overlaid to help orientate the openings on top. *Below left*: Because the remainder of the top is all one face, we have no geometry around which to create the two top bungs. The simplest way to deal with this is to add two smaller cylinders. Drawing these uses the same processes as before.

Scale	scale of mm per foot	per foot	One foot in mm	scale reduction	% Of true size
O	7	1	304.8	43.5	2.30
OO	4	1	304.8	76	1.32
N	2	1	304.8	148	0.68

Final scaling. The final operation in Blender is to scale all three parts but before doing so it is necessary to join all the parts as one, otherwise they will all scale independently and the two bungs will be left hanging in mid-air! Having done that, we now have the barrel at a full size and can therefore accurately reduce it to scale which for this example is O gauge. Looking at the following tables, we can see that the size of the model will need to be 2.3% of its original size. Other scales are included for information.

Applying the scale reduction gives measurements as follows.

Scale \ Reality	width mm	height mm
	600	900
O	13.79	20.69
OO	7.89	11.84
N	4.05	6.08

Below: **Exporting the drawings**. All the drawing programs mentioned previously will have an option for exporting the file in STL format. In Blender we select the object to export and then go into the Export menu. This is under the File menu, submenu Export, option STL. If you have multiple objects in your drawing, ensure that the 'Selection only' box is checked, as highlighted in the picture. The extracted STL format drawing can then be manipulated to bring in physical supports for the object when printing and to determine the print slices to be made to the object.

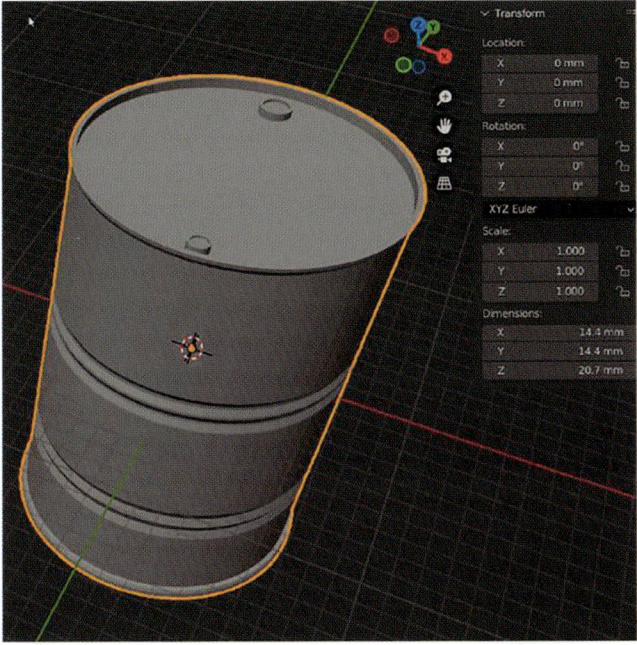

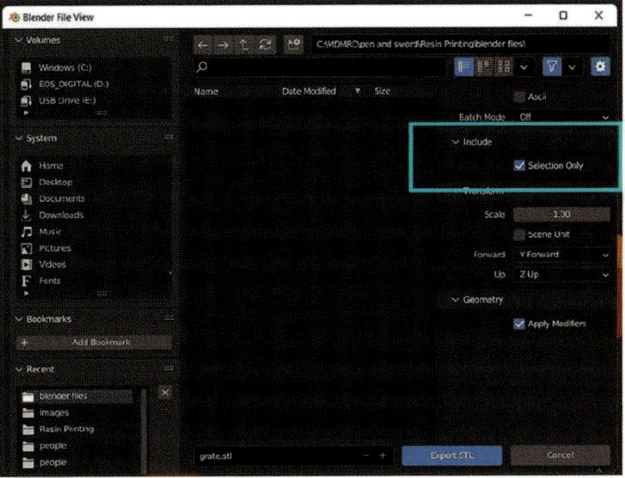

Left: The completed object ready for print extraction. This image shows the barrel and the reference image scaled down. The dimensions do not quite agree with the table due to rounding differences.

Using CHITUBOX software. Having designed the STL in Blender we have in effect a detailed 3D CAD element, but no definition of how it can go from the virtual world to the physical. CHITUBOX is the name of the software distributed with the Mars printer (and many others). In this screenshot, we see the barrel imported into CHITUBOX, ready for supporting and slicing. The Scale button on the left has been pressed to confirm that the size imported is correct.

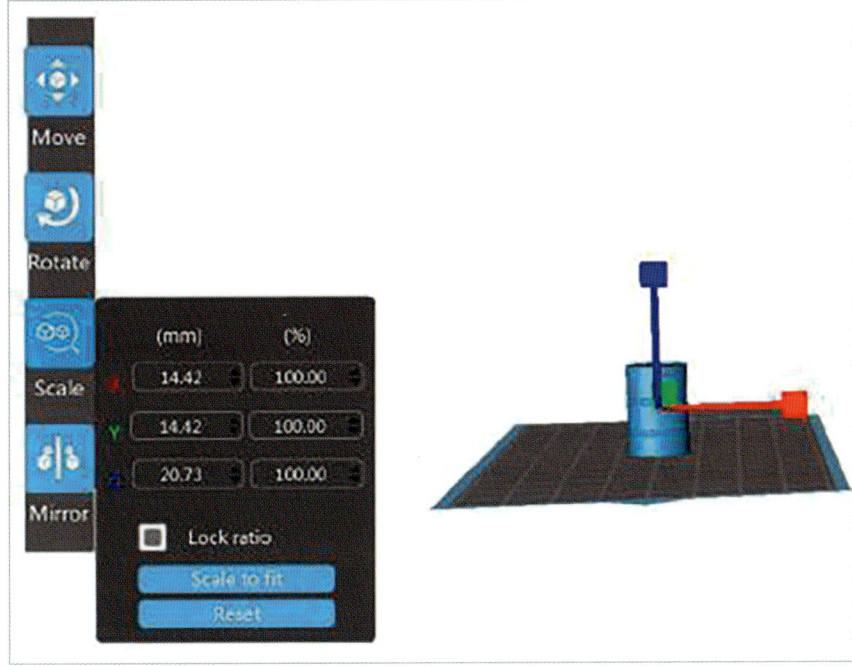

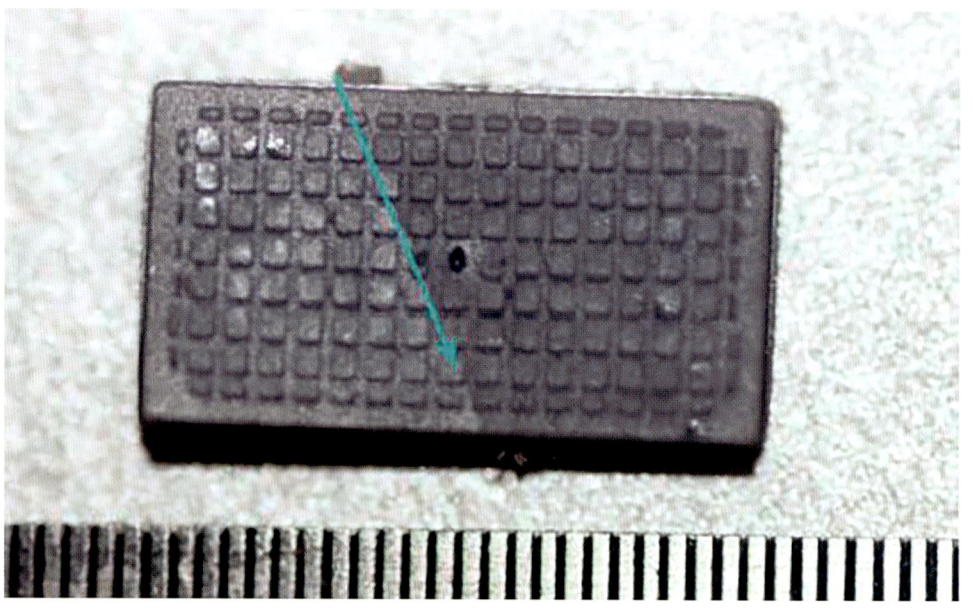

Supporting. Whole videos on YouTube cover this subject. In principle, if you print something face down on the build plate as shown here, you can expect breakages when you try to remove it. Alongside is a picture of a road grate that was printed directly on the build plate. With FDM type filament printing this would probably survive. At only 2mm thick in resin, it was too delicate to survive the parting process, fracturing along the line indicated.

Top: For an FDM print the barrel would sit flat on the print bed and be printed from the ground up. For resin printing the barrel was tipped over at 30° and supports provided. In this way it is more likely that the supports will break than the piece itself. To make room for the supports, the barrel needs raising and CHITUBOX will do this automatically once instructed that supports are required.

Middle: In CHITUBOX a look from underneath will reveal where supports are needed as indicated in purple on the base and one of the rims. Quite where and how supports are put in place can be a matter of trial and error and experience. The aim is to avoid large unsupported areas which will cause the print to fail. Some slicers will have an auto support generation feature.

Lower: 'Islands' – areas of the print that are not immediately joined to the main body of the print – should be avoided For example, if printing a figure, the hands and fingers would need separate support arrangements as they may otherwise just float away before they get attached to the body. There are no islands in this case and a suggested set of supports is shown here. Supports have been minimised since each one may leave a small dimple which will need fettling. They are also positioned so that they are on just one face of the barrel.

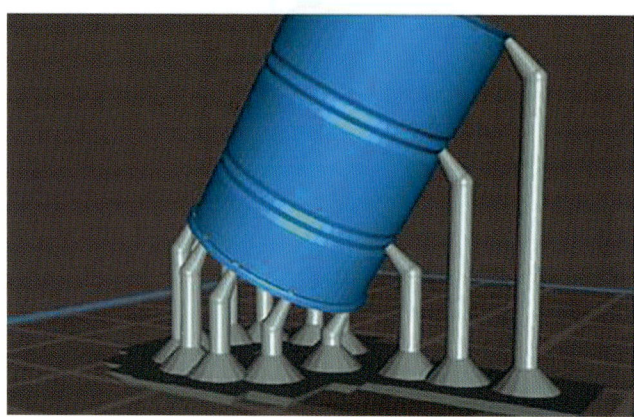

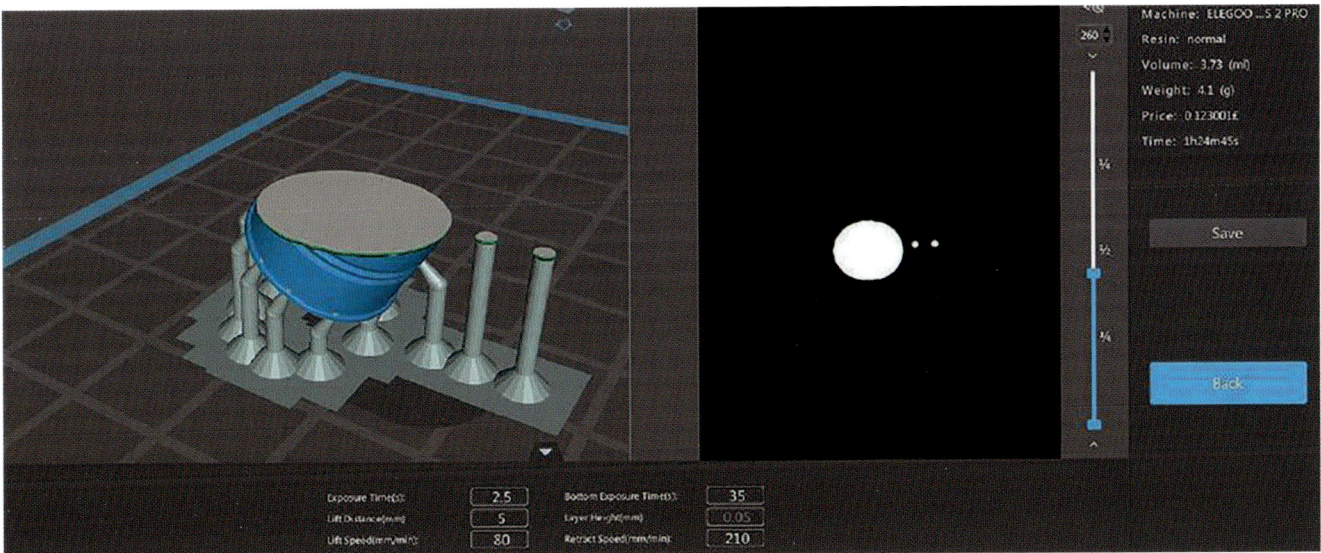

Slicing. Slicing is the term applied when you ask the software to calculate which part of your screen needs to be turned on for each layer, or slice, of your print. The software generates in memory a file to tell the printer where to expose on each layer. In basic terms this can be thought of as a file containing a .PNG format picture of each layer. The .PNG format allows transparency whereas others, like .JPG do not.

- At this stage, it is good practice to review the print process, checking particularly for unsupported islands. CHITUBOX has a slicer control on the right to enable you 'play back' each layer of the print so that you can see the printing process almost as an animation. The picture shows what is being printed on layer 260 which just happens to be part of the body of the drum and on the right, two dots that could be islands but are in fact two of the supports for higher up the barrel body. The printing time is shown on the right and is given as 1 hour, 24 minutes and 45 seconds. One of the differences between FDM and resin printers that is often not appreciated by beginners is that printing one or more objects makes no difference to the print time as all that happens is that the LCD screen below the FEP simply turns on pixels, whereas on an FDM printer, the print head must go round each object layer by layer. As a rough guide, if printing ten barrels on an FDM printer, it would take ten times as long. For that reason, on a resin printer, after having done a test print, it is common practice to print multiple copies, taking advantage of this efficiency.

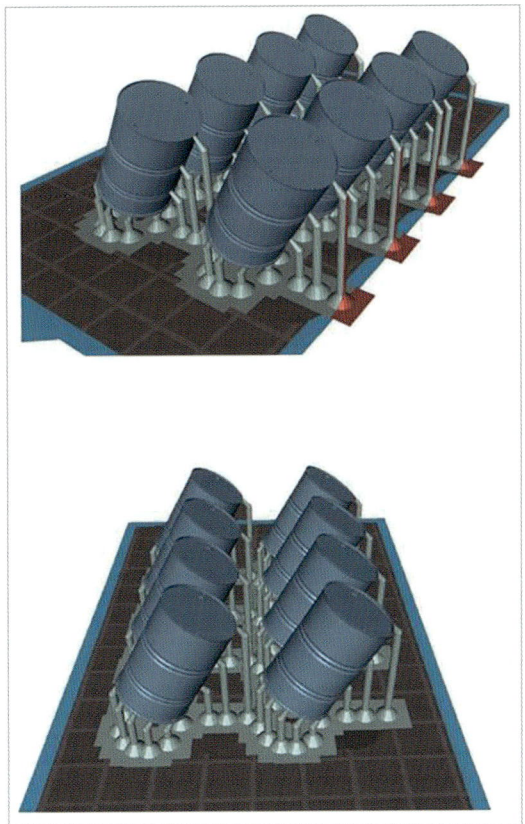

Here, the use of the Clone button has quickly added seven more barrels which have the same supports as the original. Note that on the right, four of the support base plates overhang the edge of the build plate as indicated in red. You will need to move the barrels manually or, as in this case, have them rearranged automatically if your software has this facility.

Until we meet again in the fourth book, covering the Club modelling of, and the history of, Euston Station.

Above left and above right: Mixing your hobbies. Once you have mastered either type of 3D printing, you can use the ability to remaster models using design software to blend technical ideas. Shown below is an experiment in 7mm O gauge by Brian. Making use of a 3D resin printed figure with a hollow cylinder defined in the left hand. Through this a steel rod goes under the board to a servo that is controlled by a programmed Arduino open-source electronics platform. This can set variable times and trigger events. So, the stop/go man rotates his board to allow passage of traffic.

Why We 'Do' Model Railways

Stratford built LNER Class J70 (GER C53) 0-6-0T tram 68216 trundles through St Peter's Dock, Ipswich, on 19 July 1952. The locomotive was withdrawn the following year. Model railway clubs construct layouts from the diminutive through to the monumental. If it captures the popular imagination of the members a project can be born from a single evocative image. (*Online Transport Archive –Meredith 272-4*)

We get the opportunity to take our layouts to shows to be shared with the public. Here O gauge Butterwick, which has appeared in all three books, gets an airing in King's Lynn. Stock cassettes, traverser and board extension projects all in place. (*Author*)

Micro teams of old and new members get together over layouts. Here Amberdale gets refurbished and made ready to exhibit. Everybody learns new skills. In the days of rural or age-imposed isolation, 'men's sheds' and model clubs are a great way to get together. (*Author*)

You get to support the local community. Here a Market Deeping Christmas event has Springbluff City 1957 in N as a focus. (*Author*)

You get to share hidden works of art. A scratch-built O gauge LNER consist by Brian Bartholomew runs on the Club test track. With youth sections, clubs can impart an apprenticeship of older skills and learn new ones from the fresh minds. (*Author*)